Praise for

THE LOST TOMB

"[S]o skillfully sketched is the lure of the unknown in Preston's collection of essays. From the safe distance of the pages of *The Lost Tomb*, we are allowed a delicious taste of what it is to be consumed with the desire to know, even when all evidence points to the fact that, maybe, we are better off leaving a mystery alone." —*Bookpage*, starred review

"[A] gripping compendium of his journalistic work...Throughout, Preston tackles his subjects with the obsessive enthusiasm of an amateur detective and the skills of a seasoned novelist; even those who read the articles when they were first published will take pleasure in new afterwords that provide updates about Preston's theories. This is unbeatable reading for armchair sleuths." —*Publishers Weekly*, starred review

"Buffs of buried-treasure and long-ago true-crime tales will enjoy Preston's expertly woven tales." —*Kirkus Reviews*

"The pieces are so good and the reporting so thorough that *The Lost Tomb* is a worthy addition to library collections." —*Booklist*

"*The Lost Tomb* is a seminal work about things that really do go bump in the night that proceeds to show us exactly how that can be. An extraordinary achievement in all respects." —BookTrib

"If Indiana Jones had kept a diary, it might have read much like Preston's new nonfiction work. After discovering a forgotten city in the Amazonian forest, an undertaking he documents in *The Lost City of the Monkey God*, the author travels across the globe, seeking to solve some of the ancient world's most beguiling mysteries." —Alta

"*The Lost Tomb* could not have come along at a better time...this is a book worth savoring." —*Book Reporter*

"[A] must-read for adventure, true crime, and New Mexico enthusiasts." —*New Mexico Magazine*

"Anyone who loves to peer behind the curtain and see what secrets have been hidden by history will be thrilled by the very true-to-life tales that fill this book." —*Werd*

Praise for

THE LOST CITY OF THE MONKEY GOD

A #1* New York Times *and #1* Wall Street Journal *Bestseller

"Memoirs of jungle adventures too often devolve into lurid catalogs of hardships [but] Preston proves too thoughtful an observer and too skilled a storyteller to settle for churning out danger porn. He has instead created something nuanced and sublime: a warm and geeky paean to the revelatory power of archaeology...Few other writers possess such heartfelt appreciation for the ways in which artifacts can yield the stories of who we are." —*New York Times Book Review*

"A well-documented and engaging read...The author's narrative is rife with jungle derring-do and the myriad dangers of the chase." —*USA Today*

"Deadly snakes, flesh-eating parasites, and some of the most forbidding jungle terrain on earth were not enough to deter Douglas Preston from a great story." —*Boston Globe*

"Breezy, colloquial, and sometimes very funny...A very entertaining book." —*Wall Street Journal*

"This modern-day archaeological adventure and medical mystery reads as rapidly as a well-paced novel, but it is a heart-pounding true story."
—*Shelf Awareness,* starred review

Praise for
THE MONSTER OF FLORENCE

One of USA Today's *Top True-Crime Books of All Time*

"A most unconventional thriller, a real-life murder mystery in which the authors become the suspects...Fascinating." —Associated Press

"Preston's account of the crimes is lucid and mesmerizing."
—*TIME*

"As taut and tense as any of the author's bestselling thrillers...Fascinating, stomach-churning...nerve-tingling action and vivid writing...*The Monster of Florence* is a gripping tale, filled with shocking crimes, boldly drawn characters, and the careening suspense of the ultimate whodunit."
—*Dallas Morning News*

"One of the best true-crime mysteries I've ever read...Nonfiction doesn't get any better than this." —*Kansas City Star*

"Remarkable true-crime story...passionately describes the investigations gone wrong...Preston knows how to load his storytelling with intriguing evidence and damning details. His feverish style keeps the reader turning pages with the hope of uncovering the killer's identity. In the book's most chilling moment, Preston and Spezi come face-to-face with their most likely suspect." —*USA Today*

"This bit of real-life Florence bloodletting makes you sweat and think, and it presses relentlessly on the nerves."
—*Publishers Weekly,* starred review

NONFICTION ALSO BY DOUGLAS PRESTON

The Lost City of the Monkey God

The Monster of Florence
(with Mario Spezi)

The Black Place
(with photographs by Walter W. Nelson)

The Royal Road

Talking to the Ground

Cities of Gold

Dinosaurs in the Attic

THE LOST TOMB

AND OTHER REAL-LIFE STORIES OF BONES, BURIALS, AND MURDER

DOUGLAS PRESTON

FOREWORD BY DAVID GRANN

GRAND CENTRAL

New York Boston

TO
OTTO PENZLER

Copyright © 2023 by Splendide Mendax, Inc.
Foreword copyright © 2023 by David Grann
Cover design by Eric Fuentecilla.
Cover photos by Elaine Thompson/Associated Press (skull);
DEA / M. SEEMULLER / Getty Images (map).
Cover copyright © 2025 by Hachette Book Group, Inc.

Hachette Book Group supports the right to free expression and the value of copyright. The purpose of copyright is to encourage writers and artists to produce the creative works that enrich our culture.

The scanning, uploading, and distribution of this book without permission is a theft of the author's intellectual property. If you would like permission to use material from the book (other than for review purposes), please contact permissions@hbgusa.com. Thank you for your support of the author's rights.

Grand Central Publishing
Hachette Book Group
1290 Avenue of the Americas, New York, NY 10104
grandcentralpublishing.com
twitter.com/grandcentralpub

First published in hardcover and ebook in December 2023
First Trade Paperback Edition: January 2025

Grand Central Publishing is a division of Hachette Book Group, Inc. The Grand Central Publishing name and logo is a trademark of Hachette Book Group, Inc.

The publisher is not responsible for websites (or their content) that are not owned by the publisher.

The Hachette Speakers Bureau provides a wide range of authors for speaking events. To find out more, go to hachettespeakersbureau.com or email HachetteSpeakers@hbgusa.com.

Grand Central Publishing books may be purchased in bulk for business, educational, or promotional use. For information, please contact your local bookseller or the Hachette Book Group Special Markets Department at special.markets@hbgusa.com.

Print book interior design by Jeff Williams.

Library of Congress cataloging record available at https://lccn.loc.gov/2023026359

ISBNs: 978-1-5387-4123-8 (trade paperback), 978-1-5387-4124-5 (ebook)

Printed in the United States of America

LSC-C

Printing 1, 2024

CONTENTS

Foreword by David Grann   *xi*

Introduction: Origin Stories   *xiii*

UNCOMMON MURDERS

A BURIED TREASURE   3

I went searching for a long-lost friend and a treasure we once buried. But when information is everywhere, some things are better left hidden.

THE MONSTER OF FLORENCE   11

My wife and I had always dreamed of living in Tuscany. Then we discovered that the lovely olive grove outside our door had been the site of one of the most horrific murders in Italian history.

UNEXPLAINED DEATHS

THE SKELETONS AT THE LAKE   45

The genetic analysis of hundreds of human skeletons found at a lake high in the Himalayas raises questions about why they were there, what killed them—and above all, who they were.

THE SKIERS AT DEAD MOUNTAIN — 63

The outlandish deaths of a group of skiers in the Ural Mountains is one of the most baffling mysteries from the Soviet era in Russia.

THE SKELETON ON THE RIVERBANK — 79

Who was Kennewick Man and where did he come from—and why is the government withholding his bones from scientists?

UNSOLVED MYSTERIES

THE MYSTERY OF OAK ISLAND — 105

The so-called Money Pit on Oak Island, Nova Scotia, has been the center of treasure-hunting activity for over two hundred years. The search has claimed many lives and fortunes. What could be buried there, and who buried it?

THE MYSTERY OF SANDIA CAVE — 117

Archaeologists believed that something wasn't quite right with twenty-thousand-year-old Sandia Man, whose discovery by the legendary Frank Hibben revolutionized American archaeology.

THE MYSTERY OF HELL CREEK — 151

A young paleontologist discovered an extraordinary record of the deadliest event in the history of life on Earth.

CURIOUS CRIMES

THE CLOVIS POINT CON — 181

Nobody knew how the ancient Clovis mammoth hunters created their extraordinary weapons, except for one man—and nobody believed him.

TRIAL BY FURY — 197

For years the Internet blazed with hatred toward Amanda Knox, accused and ultimately acquitted of murdering her roommate in Perugia, Italy. What really happened, and why was the Internet so angry?

OLD BONES

SKELETONS IN THE CLOSET — 223

Museum storage rooms are packed with stolen skeletons and disinterred burials. Native Americans want them back.

CANNIBALS OF THE CANYON — 239

A controversial archaeologist uncovers the dark and terrible truth about a great prehistoric civilization of the American Southwest.

THE LOST TOMB — 265

An archaeologist makes the biggest find in Egypt since King Tut's tomb: the enormous, long-lost sepulchre containing the many sons of Ramesses the Great.

AN EXTRA ADVENTURE

IN SEARCH OF THE SEVEN CITIES OF GOLD — 297

In 1540, the Spanish conquistador Coronado led an expedition into the American Southwest on a violent and failed search for the legendary Seven Cities of Gold. Four hundred and fifty years later, a friend and I retraced on horseback a thousand miles of his route across Arizona and New Mexico—nearly killing ourselves in the process.

FOREWORD BY
DAVID GRANN

THE FIRST THING you notice about these remarkable true tales by Douglas Preston is that they all contain elements of intrigue. There is a story about the unexplained deaths of a group of skiers in the Ural Mountains, and another about a hunt for treasure that has consumed seekers for nearly two centuries, costing millions of dollars and killing half a dozen people. Still another tale explores one of the most harrowing cases in the annals of crime—a string of inexplicable killings in the bucolic hills of Florence, which has generated a bewildering array of suspects. Some of the mysterious incidents Preston probes reach back thousands of years, involving the fate of ancient civilizations—where they came from and why they suddenly vanished. The evidence now consists of artifacts and bones.

Not only are the subjects in this book fascinating, but so are the investigators. They include daring archaeologists, vindictive police detectives, renegade scientists, and obsessive amateur sleuths. They can be brilliant, and fallible. And some of them seem to have their own secrets. Are they shining a light on the truth or purposely trying to cover it up?

The second thing you notice about these tales is the way they are told. An acclaimed novelist of murder mysteries, Preston has an unerring sense of suspense, of how to hold the reader in his grip. Yet in these reported pieces, he is also scrupulous and rigorous about the facts. Propelled by his own compulsive curiosity, he follows one murky trail after another, which lead him from police interrogation rooms to pits of dinosaur bones, from DNA laboratories to Egyptian tombs. He keeps on digging and digging even when he arouses the ire of authorities or faces peril. He hunts down suspected criminals, confronting them with the damning evidence he has gathered, though he always judiciously allows them to share their side. It is not a coincidence that at least three titles in this collection contain the word "mystery." Preston is a recoverer of what is unknown: answers, justice, fragments of history.

Occasionally, he must find his way through a fog of information and disinformation. When he began searching for evidence to identify a serial killer, he was convinced that he could "find the truth." Yet eventually he confesses, "I am not so certain. Any crime novel, to be successful, must contain certain basic elements: there must be a motive; evidence; a trail of clues; and a process of discovery that leads, one way or another, to a conclusion. All novels, even *Crime and Punishment*, must come to an end. But life, I have learned, is not so tidy."

It is this untidiness—this moral complexity—that makes these stories so powerful. They shed light on the greatest riddle of all: the human condition. Preston says he has drawn on the unfathomable elements in these stories to conjure the plots of his novels. As he proves in this collection, truth really can be stranger than fiction.

INTRODUCTION:
ORIGIN STORIES

THE MOST COMMON question a novelist is asked is: Where do you get your ideas? *The Lost Tomb* is my answer. I could never have become a novelist without first being a nonfiction journalist. Many of the central ideas in my novels, and the ones I've written with Lincoln Child, originated in histories that are absolutely true, albeit often bizarre and disturbing.

My mother told us children many stories about buried treasure. It was one of her favorite subjects. At my grandparents' cottage on the coast of Maine, there was a rock on the shore, uncovered only at low tide, on which was carved a worn and barely readable inscription:

R K
1743

The rock, my mother said, had been carved by a pirate to help mark the location of a treasure he had buried on a nearby promontory called Browns Head. Rumors had long circulated of a treasure on Browns Head when, in the nineteenth century, a boy got lost while hunting and found, hidden by a clump of junipers, an iron ring with a chain attached to it that went down into the ground. He placed his red-and-black checked cap on it to mark it. But when he tried to find the spot again the next day and in the days following, he couldn't.

My brothers and I searched all over Browns Head with a metal detector looking for that ring, without success. The inscription is still there, being slowly erased by the ceaseless action of the sea.

Among other stories of buried treasure my mother told me was perhaps the most famous one of all: the Oak Island Treasure. She told it so many times I can remember how it started:

> In 1795, three boys who lived on Mahone Bay in Nova Scotia got in a skiff and rowed to an uninhabited island they had always wanted to explore. The island was covered with oak trees and was called Oak Island. There, in the center of the island, they found a mystery. An old clearing had been made in the forest, leaving one giant tree standing in the middle. On an overhanging limb they found what looked like old rope burns, and the ground below had subsided into a depression. The boys were sure a pirate treasure must have been buried under the tree, because the area was known to have once been a pirate lair, so later they returned with picks and shovels and began to dig. And this is what they found . . .

I was spellbound by the mystery of Oak Island. I read books about the treasure and the many attempts by treasure hunters to overcome the island's booby-trapped tunnels and pits and retrieve a treasure thought to be worth millions, if not billions. It is possibly the most enigmatic story in all the annals of buried treasure, and I often wondered what might be hidden in the Money Pit.

When I grew up, graduated college, and went out on my own to become a writer, I began casting around for an idea for a nonfiction book. (I was not interested in writing fiction at this point.) If there was one subject that still fascinated me, it was the Oak Island treasure, and I realized as a journalist I could investigate it for real, go to Oak Island, spend time with the treasure hunters, and get paid to do it. I did some research and found that the hunt for treasure on Oak Island was more active than ever. Part of the island had been purchased by a wealthy man in Montreal named David Tobias, who was planning a huge excavation that, he claimed, would solve the mystery once and for all. He had gathered together a syndicate of investors who were preparing to float an initial public offering on the Vancouver Stock Exchange to finance the digging of what they called the "Decisive Conclusion Shaft"—a gigantic, ten-million-dollar hole to be excavated deep into the heart of the island, which would finally determine what lay buried at the bottom of the booby-trapped treasure pit.

I proposed the idea to *Smithsonian* magazine. They loved it and sent me to Oak Island, where I spent ten days with the treasure hunters and wrote a piece for the magazine. It was one of my first assignments as a journalist. I remember standing at the lip of the neglected and abandoned Money Pit for the first time—this object of my childhood fascination—staring past ruined timbers draped with moss and weeds into its mysterious blackness, while smelling the clammy, foul air exhaling from its rotten mouth. Nearby was Borehole 10X, which the treasure hunters claimed had pierced the treasure chamber before collapsing, now the center of furious activity trying to reopen it. My article was published in *Smithsonian* in June 1988. (See the story on page 105.) A decade later, an editor at *Smithsonian* told me the article was the most popular the magazine had ever published.

Around the same time, I wrote a nonfiction book, *Dinosaurs in the Attic*, edited by a rising young star at St. Martin's Press named Lincoln Child. The book told the story of the American Museum of Natural History in New York, where I worked. One night, just for fun, I gave Linc a midnight tour of the closed and darkened museum. We found ourselves in the Hall of Late Dinosaurs, surrounded by the giant skeletons of T. rex and Triceratops, illuminated only by emergency fluorescent strips in the ceiling, the bones casting ghastly shadows. Linc turned to me and said, "Doug, this is the scariest damn building in the world. We've got to write a thriller set in this place!"

That moment led to our novel *Relic*, which was made into a movie by Paramount Pictures. Linc and I found that we loved writing thrillers together, so we wrote a second novel, *Mount Dragon*, and then a sequel to *Relic* called *Reliquary*.

After *Reliquary* was published, Linc and I cast around for a fresh idea. One evening, we were enjoying a wee dram of Macallan single malt scotch on the porch of Linc's house, when the conversation turned to the Oak Island mystery and the article I had published in *Smithsonian* some years back, which Linc had read.

"Let's write a thriller based on that Oak Island story of yours!" Linc suddenly cried in a eureka moment.

I responded that I didn't think it was a good idea. Oak Island was in Canada, not the U.S., and there was no resolution to the mystery. Nobody knew what was buried in the island, if anything, or who might have buried it.

Linc dismissed all this with an impatient wave of his hand. "We're novelists. We can make up whatever we want!" He went on, "First, we move the treasure island to Maine. Didn't you once tell me about a treasure buried somewhere near your place up there? Then we can dream up answers to all those mysteries: Who buried the treasure? Why was it buried? Who engineered the pit and the booby traps? How was it done? And of course—*what* is buried there? Doug, we can work out our own solutions to all those problems! The novel will be about a fabulous treasure hunt, where everything goes to hell and people die and the big reveal—*what* is buried—is totally insane and unexpected. Think about it!"

I was dubious at first about extracting fiction from nonfiction. Would this somehow diminish the novel? Would basing it on a true story make it look like we were plagiarizing reality? But as we proceeded to toss around ideas, and as we got deeper into the bottle of Macallan, eventually we answered all those questions with what I modestly like to think were original and even brilliant ideas. Linc pointed out that almost all great writers, from Shakespeare on down, based their work on true history.

Out of this conversation, our novel *Riptide* was born. (Naturally, there's a central character in it named Macallan.) It was fun and rewarding transforming the true Oak Island, Canada, mystery into a fictional Ragged Island, Maine, treasure hunt. We had a rich source of material to mine from my article, interviews, and research, but we also had a large canvas on which to paint our own ideas. *Riptide*, too, was picked up by a studio to make a movie, but, alas, the movie was never made.

Since then, many if not most of our ideas for novels have sprung from nonfiction stories, particularly pieces I've written for the *New Yorker*.

After *Riptide*, we wrote a fifth novel, *Thunderhead*, that deals with, among other things, prehistoric cannibalism in the American Southwest. *Thunderhead* sprang directly from my article in the *New Yorker* "Cannibals of the Canyon" (page 239). And again it was Linc who helped figure out how to transform the magazine article into a novel. The latest example of this is my *New Yorker* piece called "The Skiers at Dead Mountain" (page 63). This story explored the Dyatlov Pass incident, the apparently inexplicable mass death of skiers in the Ural Mountains of Russia in 1959. Nine cross-country skiers were camped on a remote mountain in the dead of winter. Something so terrifying happened in the middle of the night that they felt compelled to cut their way out of the side of the tent and flee, half dressed and mostly barefoot, into a blizzard in twenty-below-zero weather. They fled a mile into a forest, where they froze to death. Some bodies were found with inexplicable injuries, and several were wearing clothing that tested positive for radioactivity. The unsolved mystery of Dyatlov Pass, like the unsolved Oak Island mystery, was behind the idea for the Nora Kelly novel we published in August 2023, entitled *Dead Mountain*. As we did with *Riptide*, we moved the setting—from the Urals to the Manzano Mountains of New Mexico—with a new cast of American characters. Also like *Riptide*, we imagined a shocking and original solution to the mystery.

There are many other instances of a nonfiction-fiction connection in my work. My novel *Tyrannosaur Canyon* involved the discovery of a dinosaur killed by the asteroid impact that caused the mass extinction at the end of the Cretaceous Age. That novel was read by a young graduate student of paleontology named Robert DePalma, who sent me a fan email. A few years later, he called me to say that he'd made more or less the same discovery I'd speculated about in my novel: he had found not just a dinosaur but an entire graveyard of animals killed by the asteroid impact. That call led to my *New Yorker* piece "The Mystery of Hell Creek" (page 151), which chronicled DePalma's real-life discovery of one of the most important fossil sites ever found, in the Hell Creek geological formation of North Dakota.

I've long believed that a love of stories is embedded in our very genes. We evolved to crave storytelling. Stories tell us who we are; they transmit cultural values across the generations and provide society with stability and continuity. Every society has a treasured collection of sacred stories that convey its history, values, and spiritual ideals. But good stories are not lectures on life and morals—they tell of real people engaged in dramatic events, experiencing danger and crisis, pushed to the limit—tales full of tension, excitement, good and evil, heroism and perfidy. Above all, a good story must have an arc that helps the reader make moral sense of what is—let us be frank—an arbitrary, bewildering, cruel, and fearsome world.

The Lost Tomb is a gathering of the origin stories of some of my most important novels. I want to emphasize, however, that all the stories in this book are absolutely true and have been meticulously vetted and confirmed, especially by the *New Yorker* magazine, whose fact-checking department is legendary. There is not one word of fiction in this collection, even though some of the stories in the book will strike you as being as crazy and improbable as any of my thrillers. As Dean Koontz once observed, "We craft fiction to match our sense of how things ought to be, but truth cannot be crafted. Truth *is*, and truth has a way of astonishing us to our knees, reminding us that the universe does not exist to fulfill our expectations."

I hope this collection of true stories astonishes you to your knees.

UNCOMMON MURDERS

A BURIED TREASURE

SOME WRITERS DRANK when the words didn't come. Now we have the Internet. Whenever I get stuck writing, instead of sliding open the bottom drawer with the whiskey bottle, I load up the *New York Times* or *Politico*, check my email, or, when all else fails, start Googling old acquaintances. Most of us have done it. Whatever happened to that bucktoothed kid from third grade? There he is, grinning at you from the computer screen—balding, paunchy, mustachioed—and you're reading about his lake cottage, his wood-turning hobby, his Danube cruise, his grandchildren, his cats.

A few months ago, as I was staring at a wretched chapter I was trying to write, I idly Googled the name "Peter Anderson" and "New Jersey." Petey was my best friend growing up in Wellesley, Massachusetts, until he moved to New Jersey in seventh grade. But his was a common name and it returned tens of thousands of hits. Slumped in my chair, continuing to waste time, I tried his mother's name, his father's, and his brother's. There were just too many Andersons, though, and nothing of note surfaced, beyond an old article in the *Times* of Trenton about a murder. This obviously wasn't my friend Petey: there were probably dozens of other Peter Andersons in New Jersey, alive and well and going about their business.

Petey lived across the brook from me in a white stucco house overlooking a golf course owned by Wellesley College. He was a droll kid

---

Originally published in *Wired* magazine in 2019 as "The Lost Trail."

with pale orange hair and papery skin through which you could see blue veins. He had a cheerful mother and a silent, raddle-faced alcoholic father. After work, his father parked himself in a wing chair in the living room, shook out the afternoon *Boston Herald*, and read it while gripping a scotch on the rocks. When he wanted another, he shook the empty glass and Mrs. Anderson hurried in with fresh ice and the bottle.

In those days before the curated childhood, Petey and I ran wild, concerning ourselves with knocking on doors and running away, getting chased off the golf course by greenskeepers, playing stickball, making crank calls, and hunting for buried treasure. We dug holes in the woods behind the golf course, hoping to unearth a sack of pine tree shillings from colonial days or gold doubloons from the time when Captain Kidd (so we fantasized) sailed his ship up the Charles River.

One fall day, my mother gave me an empty cookie tin with a picture of a great sailing ship plowing through waves, surrounded by gulls. Petey came over and I said, "Let's fill this with treasure and bury it." We decided to leave it in the ground for ten years and dig it up when we were eighteen. The year was 1964.

Petey and I spent hours debating what to put in the tin. The treasure had to be something valuable enough that our grown-up selves would be glad to have it back. We gathered our best things and laid them out on my bed for inspection. Most of them struck us as childish crap, but a few stood out as objects with adult gravitas. I chose a Morgan silver dollar, a coiled-up trilobite fossil, and my finest arrowhead—an ancient beauty flaked out of petrified wood in which you could still see the tree rings. Among Petey's treasures were a squirrel skull, a miniature brass cannon from the USS *Constitution*'s gift shop, and an intricate blob of lead he had made by melting fishing sinkers on the stove and pouring the molten metal into water. It was a method of telling the future, he said. The blob predicted that his life would be one of wealth, success, and happiness.

As we looked over our carefully assembled treasures, they still didn't seem adequate for a great journey into the future. I had an idea: Why not each write the story of our lives? Whatever else we put in the tin, we knew this would make for good reading, especially if we'd forgotten our childhoods, like most adults we knew.

Every afternoon for several weeks, we gathered in Petey's living room and labored over legal pads, sharpening and resharpening our pencils until the flakes lay in tiny curls all around us on the carpet. I called my opus "The Story of My Life So Far." I remember writing about riding horses in Wyoming and living with a Luo family in the African bush, eating ram guts and ugali for supper and listening to the lions roar at night. (I had eccentric parents.) Petey called his "Eight Years Old and Full of Beans," a title I criticized as corny, but he stuck with it. I didn't read his story and he didn't read mine; that would happen ten years hence.

When we were done, we rolled up the papers, tied them with ribbons, and sealed them with wax. We carefully packed the tin, then wrapped it in layers and layers of duct tape so it would be waterproof. I wondered: When we dug it up, who would we be? What would America look like? Would there be flying cars and men on the moon? Would we all be Reds, as my teacher said was going to happen if people didn't wake up? Would the world be a cinder from nuclear war? The future was strange, scary, and thrilling to ponder.

The crucial question arose as to where to bury the tin. It had to be a far-off place where anyone could find it—that was part of the excitement. We settled on an abandoned field deep in the woods at the farthest edge of the Wellesley College property. One fine fall day, equipped with a compass, pick, and shovel, Petey and I set out. The maples had turned scarlet and the leaves, backlit by the sun, glowed like church glass against a blue sky.

Once at the field, we set down the shovel and pick and scoped out the scene. We settled on a hollow oak as our starting point. Standing at the foot of the oak, I sighted into the field with my compass. Due west stood a cedar sapling in the middle of the field. We paced the distance through weeds and grass, picking the cockleburs off our socks: twenty-one paces. Having established that distance, we then measured nine paces due north.

I sank the shovel into the ground and cut a rectangle in the tough grass. We worked off the turf in a single piece and laid it aside. Wielding the pick, I broke up the dirt while Petey shoveled it out, piling it nearby. In twenty minutes we had prepared a most excellent hole, two feet deep, which cut through the upper layer of loam into a stratum of orange clay. We placed the tin inside and snugged it down tight. Petey backfilled the hole and smacked down the dirt with the back of the shovel. We replanted the grass on top, brushed it out with our fingers, and threw the remaining clods of dirt into the woods. A dozen artfully placed autumn leaves completed the picture.

We shook hands and promised to come back in ten years. I drew a treasure map showing the oak tree, the field, and the cedar tree, with dotted lines indicating the number of paces and directions, leading to where an X marked the spot. I made a copy of the map for Petey and locked mine in a tin safe hidden behind a secret panel in my room, and there it remained as the years rolled by.

In seventh grade, Petey moved away to New Jersey. We were both devastated by the separation. We wrote long letters to each other, some so fat they had to be rolled up and mailed in a tube. But over the course of a year, the letters lost weight and became sickly, and finally our friendship passed away peacefully in its sleep. The treasure was almost—but not quite—forgotten.

When I was sixteen, while rummaging through some old stuff, I found the treasure map still hidden in the safe. I stared at it, thinking of my long-lost pal. We had promised to wait until 1974, but I hadn't seen Petey in years, and it felt like an eternity had passed since we buried the tin. The years between eight and sixteen are mighty indeed. Since Petey had left, I figured the promises we had made to each other were no longer valid, and I would be justified in digging up the box by myself, two years before its time.

I got a shovel out of the garage and, map in hand, set off. My days of roaming the woods were over, but I still knew the land by heart. When I arrived at the location of the field, however, I was shocked to discover it no longer existed. It had rejoined the forest, growing into a thicket of red cedars, oaks, and birches.

I found the hollow oak. But when I stood at its base and trained my compass west, I couldn't identify the cedar tree. There were dozens of cedars of varying sizes, some ten feet tall or more, growing in a dense stand. I measured out twenty-one paces, trying to approximate the length of an eight-year-old's stride. Pushing through the branches, I found a tree that, because it seemed to be the biggest, might have been the one we used as a landmark. From there, following the compass, I measured nine paces north and halted among a cluster of smaller trees and bushes, where I began to dig. It was hard going, hacking through mats of crisscrossing roots. After a while I hit softer dirt and the digging got easier. Down, down I dug, well into the orange layer, finding nothing. I paced it off again and dug another hole. Nothing. I dug a third. And a fourth. Resentment mingled with a keen sense of loss. Up to that point in my life, time had given me everything I had; I was now starting to realize it also took things away.

OVER THE YEARS, thoughts of Petey drifted through my mind at random moments. An old friend said he'd heard Petey was cleaning swimming pools somewhere, and that he was gay, but I didn't get around to digging deeper. That's where things would have remained, but for the constant temptation of the Internet. Now, at sixty-two years of age, I was frivolously Googling away, trying to find out what someone I knew more than half a century ago was up to, and having no success. And then I remembered a crucial detail: Petey's middle name was Stark. This was enough to refine the search—and up popped that same article in the *Times* of Trenton.

On May 2, 2011, the newspaper said, the body of a Peter Anderson was found in a boarding house in Ewing, New Jersey. The man's hands and feet were bound with packing tape. He had apparently been bludgeoned to death with a hammer. Curiously, Petey's middle name appeared nowhere in the article. It seemed that the search engine had turned over an arbitrary stone. Still, I was deeply troubled. I couldn't seem to verify that the murder victim *wasn't* Petey. Realizing I wasn't going to get any work done until I knew for sure, I spent $9.95 on an Intelius search of public records.

It revealed that Peter Stark Anderson, son of Virginia and Perry Anderson, of Hightstown, New Jersey, was deceased, and that the date of his death was May 2, 2011. The murder victim was my friend Petey.

I felt as though I had been suddenly poisoned; my insides seized up. Gripped by a feeling of nausea, I began reading the other stories about the homicide, of which there were half a dozen.

The police had quickly identified a suspect, a man named Robert Horrocks Jr., who had been working as a handyman at the boarding house where Petey lived. Horrocks had fled to Connecticut the night of the killing. His bloodstained clothes were recovered, and he was brought back to New Jersey and charged with murder. At his bail reduction hearing, his attorney, Anthony Cowell, reportedly contended that Horrocks had killed Petey while defending himself in a fight. Horrocks interrupted him and insisted on speaking. The judge, obliging, warned him of his rights. Horrocks then said, "I know I'm guilty of the crime."

At his sentencing, Horrocks went on to tell the court why he was justified in killing Petey. While doing repairs at the boarding house, he said, he had brought along his girlfriend's autistic adult son, whom Petey had sexually assaulted. I forced myself to read on, stupefied with disbelief. Horrocks returned weeks later to take revenge. "I did what I had to do," he said. "I didn't mean to kill the guy, but he ended up dying and I'm paying for it. End of story."

According to the *Times* of Trenton, the prosecutor questioned Horrocks's account. Nobody, she said, could corroborate his story of a sexual assault, including the autistic man himself. Furthermore, the prosecutor continued, Horrocks knew the son had been molested years before by another man. "The story was out there ripe for the picking, and I suspect the defendant latched onto it as a justification," she said. Horrocks was sentenced to thirty years in prison without possibility of parole.

The story left my mind in a state of complete disorder, a failure of understanding at a profound level. As a boy, Petey had been terrified of violence. Being frail and gentle, he was the target of bullying—shoulder bumping (ex-*cuse* me!), tripping, popping of shirt loops, taunts of "faggot," and smacking upside the head. He fled from any hint of conflict, usually with a wiseass comment flung over his shoulder, and he could outrun any goofus who took up the chase. I could not begin to fathom the trajectory that took him from an upper-upper-middle-class home in Wellesley—one of the wealthiest suburbs in the country—to a cramped boarding house in New Jersey. Arbitrary details of his life came flickering back into my memory: Petey singing songs to his hamster Gertrude; Petey cradling his dying dog after she'd been hit by a car, even though she was bleeding and peeing all over him; Petey writing silly stories about a magical valley where the animals talked like people; Petey and me burying a treasure.

My feckless Googling had reaped a monstrous reality that would haunt me for the rest of my life. I asked myself: Is there something righteous in facing reality, or would it have been better to stay ignorant? A surfeit of ugly knowledge is a feature of our age, a result of the Internet carrying to our doorstep, like a tomcat with a dead rat, all manner of brutal information. How many others have flippantly Googled a long-lost friend and discovered something ghastly? This was not knowledge as power; it was knowledge as sorrow.

But I was not done. I had to know what happened. I started digging. I got contact information for the prosecutor, the public defender, the judge, and the journalist who wrote the Trenton *Times* stories. I got the telephone number, address, and email address of Petey's brother, living in Massachusetts. I lined up all that information, printed it out, and squared it off on the corner of my desk. It was everything I needed to find the answer. I stared at it for weeks. And then I threw it out. In a world verminous with information, there are some things I just don't want to know.

A month or two later, when I had at least minimally processed Petey's murder, I did undertake one small bit of research. Using Google Earth, I looked at the abandoned field where we buried the treasure. It had grown into solid forest—wild, tangled, thick, a small wilderness in suburbia. Fifty-five years later, the tin containing Petey's life story, my prized arrowhead, and a lying lump of lead is still there, continuing its long, dark journey into the future.

UPDATE

After this story was published in Wired *magazine, I received an email from a journalist who covered the murder trial. She wanted to know if I'd be interested in learning more about the case. I did not respond.*

THE MONSTER OF FLORENCE

MY WIFE AND I had always dreamed of living in Italy. In the year 2000 we finally made the move with our two young children. We rented a fourteenth-century farmhouse surrounded by olive groves and vineyards in the enchanting hills south of Florence. There were two famous landmarks near us: the villa La Sfacciata, once the home of Amerigo Vespucci, the Florentine explorer who gave America its name; and the villa I Collazzi, said to have been designed by Michelangelo, where Prince Charles painted many of his watercolors of the Tuscan landscape.

The olive grove beyond our front door boasted a third landmark, of sorts. It had been the site of one of the most horrific murders in Italian history, one of a string of double homicides committed by a serial killer known as "the Monster of Florence." As an author of murder mysteries, I was more curious than dismayed. I began researching the case. It didn't take me long to realize I'd stumbled across one of the most harrowing and remarkable stories in the annals of crime.

I contrived to introduce myself to the journalist who was the acknowledged expert on the case, a former crime correspondent for *La Nazione* named Mario Spezi. We met in Caffè Ricchi, in Piazza Santo Spirito, overlooking Brunelleschi's last and greatest church. Spezi was a journalist of the old school, with a handsome if cadaverous face, salt-and-pepper hair, and a Gauloise hanging from his lip. He wore a

---

Originally published in *The Atlantic* in August 2006.

Bogart fedora and trench coat, and, knocking back one espresso after another, he told me the full story. As he spoke, he had his pocket notebook open on the table and he sketched his thoughts—I later learned it was a habit of his—the pencil cutting and darting across the paper, making arrows and circles and boxes and dotted lines, illustrating the intricate connections among the killings, the arrests, the suspects, the trials, and the many failed lines of investigation.

Between 1974 and 1985, seven couples—fourteen people in all—were murdered while making love in parked cars in the hills of Florence. The case was never solved, and it has become one of the longest and most expensive criminal investigations in Italian history. More than 100,000 men have been investigated and more than a dozen arrested, and scores of lives have been ruined by rumor and false accusations. There have been suicides, exhumations, poisonings, body parts sent by post, seances in graveyards, lawsuits, and prosecutorial vendettas. The investigation has been like a malignancy, spreading backward in time and outward in space, metastasizing to different cities and swelling into new investigations, with new judges, police, and prosecutors, more suspects, more arrests, and many more lives ruined.

It was an extraordinary story, and I would—to my sorrow—come to share Spezi's obsession with it. We became friends after that first meeting, and in the fall of 2000 we set off to find the truth. We believed we had identified the real killer. We interviewed him. But along the way we offended the wrong people, and our investigation took an unexpected turn. Spezi has just emerged from three weeks in prison, accused of complicity in the Monster of Florence killings. I have been accused of obstruction of justice, planting evidence, and being an accessory to murder. I can never return to Italy.

It all began one summer morning many years ago in the Florentine hills. The date was June 7, 1981, a Sunday. Mario Spezi, then thirty-five, was covering the crime desk at *La Nazione*, Florence's leading paper, when a call came in: a young couple had been found dead in a quiet lane in the hills south of town. Spezi, who lived in those same hills, hopped into his Citroën and drove like hell along back roads, arriving before the police.

He will never forget what he saw. The Tuscan countryside, dotted with olive groves and vineyards, lay under a sky of cobalt blue. A medieval castle, framed by cypress trees, crowned a nearby rise. The boy seemed to be sleeping in the driver's seat, his head leaning on the window. Only a little black mark on his temple, and the car window shattered by a bullet, indicated that it was a crime scene. The girl's body lay some feet behind the car, at the foot of a little embankment, amid scattered wildflowers. She had also been shot and was on her back, naked except for a gold chain, which had fallen between her lips. Her vagina had been removed with a knife.

"What shook me most of all," Spezi told me, "was the coldness of the scene. I'd seen many murder scenes before, and this wasn't like any of them." Everything was unnaturally composed, immobile, with no signs of struggle or confusion. It looked, he said, like a museum diorama.

Due to the sexual nature of the crime, it was assumed that the killer was a man. And yet the medical examiner's report noted that the killer had not sexually assaulted the woman. On the contrary, he had assiduously avoided touching her body, except to perform a mutilation so expert that the medical examiner speculated he might be a surgeon—or a butcher. The report also noted that the killer had used a peculiar knife with a special notch in it, probably a scuba knife.

Spezi's article caused a sensation: it revealed that a serial killer was stalking the countryside of Florence. In a sidebar next to the article, *La Nazione* pointed to something the police had overlooked: this killing was similar to a double homicide that had taken place in the hills north of Florence in 1974. The article prompted the police to compare the shells recovered from both crimes. They discovered that the bullets had been fired by the same gun, a .22-caliber Beretta "long barrel" firing Winchester series "H" copper-jacketed rounds, which, according to ballistics experts, probably came from the same box of fifty. The gun had a defective firing pin that left an unmistakable mark on the rim of each shell.

The investigation that followed lifted the lid off a bizarre underworld, which few Florentines realized existed in the beautiful hills surrounding their city. Because most Italians live with their parents until they marry, sex in cars is a national pastime. At night, dozens of voyeurs prowled the hills spying on people making love in parked cars. Locally, these voyeurs were called "*Indiani*," or Indians, because they crept around in the dark, some loaded down with sophisticated electronic equipment like suction-cup microphones and night-vision cameras. Following a quick investigation, the police arrested and jailed one of these Indiani.

A few months later the killer struck again, on a Saturday night with no moon, this time north of Florence, using the same Beretta and performing the same mutilation. This third double homicide panicked Florence and garnered front-page headlines across Italy.

Spezi worked nonstop for a month, filing fifty-seven articles. The excellent contacts he had developed among the police and the Carabinieri ensured he had the breaking news first. The circulation of *La Nazione* skyrocketed to the highest point in its history. Spezi wrote about one suspect, a priest, who frequented prostitutes for the thrill of shaving their pubic hair. He wrote about a psychic who spent a night in the cemetery where a victim was buried, sending and receiving messages from the dead. Spezi's articles became famous for their dry turns of phrase and that one wicked little detail that remained with readers long after their morning espresso. Florentines have a flair for conspiracy thinking, and the citizenry indulged in wild speculation. Spezi's articles were a counterpoint to the hysteria: understated and ironic in tone, they crushed one rumor after another and gently pointed the reader back to the actual evidence.

Late that November, Spezi received a journalistic prize for work he had done unrelated to the case. He was invited to Urbino to accept the prize, a kilo of the finest white Umbrian truffles. His editor allowed him to go only after he promised to file a story from Urbino. Not having anything new to write about, Spezi recounted the histories of some of the famous serial killers of the past, from Jack the Ripper to the Monster of Düsseldorf. He concluded by saying that Florence now had its very own serial killer—and there, amid the perfume of truffles, he gave the killer a name: "*il Mostro di Firenze*," the Monster of Florence.

The austere savagery of the crimes preyed heavily on Spezi's mind. He began to have nightmares and was fearful for his young and beautiful Flemish wife, Myriam, and for their baby daughter, Eleonora. The Spezis lived in a converted monastery on a hill high above the city, in the very heart of the Monster's territory. What frightened Mario most of all, I think, was that coming in contact with such barbarity had forced him to confront the existence of a kernel of absolute evil within us all. The Monster, he once told me, was more like us than we might care to admit—it was a matter of degree, not kind.

Myriam urged her husband to seek help, and finally he agreed. Instead of going to a psychiatrist, Spezi, a devout Catholic, turned to a monk who ran a small mental-health practice out of his cell in a crumbling eleventh-century Franciscan monastery. Brother Galileo Babbini was short, with Coke-bottle glasses that magnified his piercing black eyes. He was always cold, even in summer, and wore a shabby down coat beneath his brown monk's habit. He seemed to have stepped out of the Middle Ages, and yet he was a highly trained psychoanalyst with a doctorate from the University of Rome.

Brother Galileo combined psychoanalysis with mystical Christianity to counsel people recovering from devastating trauma. His methods were not gentle, and he was unyielding in his pursuit of truth. He had, Spezi told me, an almost supernatural insight into the dark side of the human soul. Spezi would see him throughout the case; he would confide to me that Brother Galileo had preserved not only his sanity but also his life.

The next killing took place eight months later, in June 1982, again on a Saturday night with no moon. The same gun was used and the same inexplicable mutilation performed. Twelve days later, an anonymous letter arrived at police headquarters in Florence. Inside was a yellowed clipping from *La Nazione* about a forgotten 1968 double murder—of a man and a woman who had been having sex in a parked car. Scrawled on the article was a bit of advice: "Take another look at this crime."

Investigators rifled through their old evidence files and found that, through a bureaucratic oversight, the shells collected in 1968 had not been disposed of. They were Winchester series "H" rounds, and each one bore on the rim the unique signature of the Monster's gun.

The police were confounded, because the 1968 murders had been solved. It was an open-and-shut case. A married woman, Barbara Locci, had gone to the movies with her lover; afterward, they had parked on a quiet lane to have sex. They were ambushed in the middle of the act and shot to death. The woman's husband, Stefano Mele, an immigrant from the island of Sardinia, was picked up the following morning; when a paraffin-glove test indicated he had recently fired a handgun, he broke down and confessed to killing his wife and her lover in a fit of jealousy. But Mele could not be the Monster of Florence: he had been in prison at the time of the 1981 killing, and had lived since his release in a halfway house in Verona.

Overnight, every crime journalist in Italy wanted to interview Stefano Mele. The priest who ran the halfway house in Verona was equally determined to keep them away. Spezi arrived with a filmmaker on the pretense of shooting a documentary on the halfway house's good work. Little by little, after taking generous footage of the priest and conducting a series of fake interviews with inmates, he reached Mele.

His first glimpse was discouraging: the Sardinian walked around in circles, taking tiny, nervous steps. An expressionless smile, frozen on his face, revealed a cemetery of rotten teeth. He mumbled rambling answers to Spezi's questions, his words defying interpretation. Then, at the end, he said something odd: "They need to figure out where that pistol is," he said. "Otherwise there will be more murders.... *They* will continue to kill.... *They* will continue."

Spezi grasped something the police would also learn: Stefano Mele had not been alone that night in 1968. It had not been a spontaneous crime of passion but a *delitto di clan*, a clan killing, in which others from Mele's Sardinian circle had participated. Investigators theorized that one of the killers had enjoyed the experience so much that he had gone on to become the Monster of Florence—using the same gun.

This stage of the investigation became known as the "*Pista Sarda*," the Sardinian Connection. It focused on three Sardinian brothers: Francesco, Salvatore, and Giovanni Vinci. All three had been lovers in turn of the woman murdered in 1968, and one or more had been present at her killing.

The police first arrested Francesco.

In September 1983, with Francesco Vinci in jail, the Monster struck again. This was the killing that took place in the olive grove beyond our front door. A German couple had parked their Volkswagen camper in the grove for the night. It was only after killing the two lovers that the Monster realized he had made a mistake: both were men, one of whom had long blond hair. Instead of performing his usual mutilation, the Monster tore up a homosexual magazine he found in the camper and scattered the pieces outside.

The authorities refused to release Francesco Vinci. They believed one of his relatives had tried to throw them off by committing a new murder using the same gun—or, at the very least, that Francesco knew who the Monster was. Investigators became suspicious of another member of the clan, Antonio Vinci, and arrested him on firearms charges. They grilled the two men relentlessly, but were unable to break them, and finally were forced to release Antonio. Francesco remained in custody.

Four months later the police electrified Florence with an announcement, and once again Spezi had the scoop. *La Nazione* carried the banner headline *"I Mostri Sono Due"*—"There Are Two Monsters." Two other members of the Sardinian group—both suspected of having been present at the 1968 clan killing—were arrested and charged with being the Monster of Florence.

Francesco Vinci was released.

All winter the police worked on the two men, desperately trying to extract confessions and develop their case—with no success. Summer arrived, and tensions rose in Florence, even though suspects were in prison. Then, in July, the Monster struck again. Again he left the empty shells, which had become, perhaps intentionally, his calling card. He mutilated the woman and, adding a new horror, amputated and carried away her left breast.

This killing, which had occurred outside Vicchio, the birthplace of Giotto, triggered a nationwide outcry and generated headlines across

Europe. Six times the Monster had attacked, killing twelve people, while the police had arrested and then been forced to release a steady stream of suspects. A special strike team was formed: the Squadra anti-Mostro, composed of both Polizia and Carabinieri. (Italy has two police forces that investigate crime, the civilian Polizia and a branch of the military known as the Carabinieri; they operate independently, and often antagonistically, especially in high-profile cases.) The government offered a reward of roughly $290,000 for information leading to the capture of the Monster, the highest bounty in Italian history. Warning posters went up, and millions of postcards were distributed to tourists entering Florence, advising them not to go into the hills at night.

For Mario Spezi, the case had become a career. His colleagues at *La Nazione* affectionately referred to him as the paper's "Monstrologer." He wrote a highly regarded book about the case that was made into two films. He often appeared on television, and his soft voice and highly developed sense of irony were not always pleasing to investigators, especially those with whom he disagreed. Spezi had a perverse passion for needling people in positions of power, and he developed a second career as a caricaturist for *La Nazione*, which regularly printed his outrageously funny cartoons of politicians, officials, and judges in the news.

At the same time, he continued to see Brother Galileo, who helped him make peace with the physical horror of the murder scenes and the metaphysical evil behind them. Galileo spent a great deal of time probing Spezi's nightmares and his childhood, forcing him to confront his own inner demons.

In the summer of 1985 the Monster resurfaced in what would be the most terrible killing of all. The victims were two young French tourists who had pitched a tent in a field on the edge of a wood, not far from the villa where Machiavelli wrote *The Prince*. According to the reconstruction of the crime, the killer approached the tent and, with the tip of a knife, made a twelve-inch cut in the fly. The campers heard the noise and unzipped the front flap to investigate. The killer was waiting for them and opened fire, hitting the woman in the face and the man in the wrist. The woman died instantly, but the man, an amateur sprinter, dashed out of the tent and sped toward the trees. The killer raced after him, intercepted him in the woods, and cut his throat, almost decapitating him. The young man's blood stained the tree branches above to a height of ten feet. The killer returned to his female victim to perform the usual ritual mutilation—and again, he carved out and carried off her left breast.

This killing occurred on either Saturday or Sunday night; the date would become a matter of the utmost importance. The bodies were discovered by a mushroom picker on Monday at two p.m. At five p.m. the police took a detailed series of photographs, which showed the bodies covered with centimeter-long blowfly larvae.

On Tuesday, one of the prosecutors in the case, Silvia Della Monica, received an envelope in the mail. It had been addressed like a ransom note, with letters cut out of magazines, and inside was the victim's left nipple. As with everything else, the killer had been careful not to leave fingerprints; he had even avoided sealing the letter with his tongue. The experience shattered Della Monica: she withdrew from the case and, soon after, abandoned her career in law enforcement.

This, so far as we know, was the Monster's last killing. Over eleven years, fourteen lovers had been shot with the same gun. But the investigation had hardly begun. A judicial storm was mounting that would change its course and perhaps guarantee that the truth would never be known—and the killer never found. There were two key players in the coming storm: the chief prosecutor in the case, Pier Luigi Vigna, and the examining magistrate, Mario Rotella.

Vigna was already a celebrity in Italy when he assumed his role in the Monster case. He had ended a plague of kidnapping for ransom in Tuscany with a simple method: when a person was kidnapped, the state immediately froze the family's bank accounts. Vigna refused to travel with bodyguards, and he listed his name in the telephone book and on his doorbell, a gesture of defiance that Italians found admirable. The press ate up his pithy quotes and dry witticisms. He dressed like a true Florentine, in smartly cut suits and natty ties, and, in a country where a pretty face means a great deal, he was exceptionally good-looking, with finely cut features, crisp blue eyes, and a knowing smile.

Mario Rotella, the examining magistrate, was from the south of Italy, an immediate cause for suspicion among Tuscans. He sported an old-fashioned mustache, which made him look more like a greengrocer than a judge. And he was a pedant and a bore. He didn't like to mingle with journalists and, when cornered, answered their questions with unquotable circumlocutions. Under the Italian system, the prosecutor and the examining magistrate work together. But Vigna and Rotella disliked each other and disagreed on the direction the investigation should take.

The two suspects had been in jail when the French tourists were killed, and Vigna wanted to release them. Judge Rotella refused. He remained convinced that one of the clan members was the Monster—and that the others knew it. For a while Rotella prevailed. His focus turned to Salvatore Vinci, who had been involved with Barbara Locci and Stefano Mele in an elaborate sexual threesome, and who appeared to have been the prime shooter in the 1968 killing. Salvatore had been forced to leave Sardinia after his nineteen-year-old wife, Barbarina, was found asphyxiated by gas in their home. The death, in 1961, had officially been determined a suicide—although everyone in town believed it was a murder. Someone had mysteriously rescued their one-year-old son, Antonio, from the gas while leaving the boy's mother to die. Rotella didn't have enough evidence to arrest Salvatore Vinci for being the Monster, so he had him arrested instead for the murder of Barbarina. His plan was to convict him for that murder and leverage it against him to identify the Monster.

The trial was a disaster: witnesses were vague, and evidence had gone stale. Antonio Vinci refused to testify against his father, at whom he glowered silently in court. Salvatore was acquitted, walked out of the courtroom, and vanished, slipping through the hands of the police, apparently forever.

This was the last straw for Vigna. He felt that the Sardinian investigation had led nowhere and brought nothing but humiliation. There was enormous public pressure to make a radical break. Vigna argued that the gun and bullets must have passed out of the hands of the Sardinians before the Monster killings had begun. He demanded that the investigation be started afresh.

Rotella refused. He was supported by the Carabinieri, Vigna by the Polizia.

It was an ugly fight, and, as is usual in Italy, it devolved into a personality contest, which Rotella naturally lost. The Sardinian Connection was formally closed, and the suspects—including the men who had participated in the 1968 killing—were officially absolved. The problem was that, if Rotella was right, the investigation could now proceed in every direction except the correct one. Officers in the Carabinieri were so angry at this turn of events that they withdrew the organization from the Squadra anti-Mostro and renounced all involvement in the case.

Vigna reorganized the Squadra anti-Mostro into an all-Polizia force under the leadership of Commissario (Chief Inspector) Ruggero Perugini, later fictionalized as Chief Inspector Rinaldo Pazzi in Thomas Harris's novel *Hannibal*. Harris had followed the case while writing the novel, and he had been a guest in Perugini's home. (The chief inspector was not altogether pleased to see his alter ego gutted and hung from the Palazzo Vecchio by Hannibal Lecter.) Perugini was more dignified than his sweaty and troubled fictional counterpart in the movie version of *Hannibal*. He spoke with a Roman accent, but his movements and dress, and the elegant way he handled his pipe, made him seem more English than Italian.

The new chief inspector became an instant celebrity when, on a popular news program, he fixed his Ray-Bans on the camera and spoke directly to the Monster in firm but not unsympathetic tones. "People call you a monster, a maniac, a beast," he said. "But I believe I have come to know and understand you better." He urged the Monster to give himself up. "We are here to help you," he said.

Inspector Perugini wiped the slate clean. He started with the axiom that the gun and bullets had somehow passed out of the hands of the Sardinians, and that the Monster was unconnected to the 1968 clan killing. The forensic examination of the crime scenes had been spectacularly incompetent: people came and went, picking up shells, taking pictures, throwing their cigarette butts on the ground. What forensic evidence was collected—a knee print, a bloody rag, a partial fingerprint—was never properly analyzed, and, infuriatingly, some had been allowed to spoil. Perugini viewed this evidence with skepticism; he was smitten by the idea of solving the crime with computers.

He examined tens of thousands of men in Tuscany, punching in various criteria—convictions for sex crimes, propensity for violence, past prison sentences—and winnowed down the results. The search eventually fingered a sixty-nine-year-old Tuscan farmer named Pietro Pacciani, an alcoholic brute of a man with thick arms and a short, blunt body who had been convicted of sexually assaulting his daughters. His prison sentence coincided with the gap in killings between 1974 and 1981. And he was violent: in 1951 he had bashed in the head of a traveling salesman whom he had caught seducing his fiancée, and then raped her next to the dead man's body.

Inspector Perugini had his suspect; all that remained was to gather evidence. In reviewing Pacciani's old crimes, Perugini was struck by something: Pacciani had told the police he'd gone crazy when he had seen his fiancée uncover her left breast for her seducer. This statement, he felt, linked Pacciani to the Monster, who had amputated the left breast of two of his victims.

Perugini searched Pacciani's house and came up with incriminating evidence. Prime among this was a reproduction of Botticelli's *Primavera*, the famous painting in the Uffizi Gallery that depicts (in part) a pagan nymph with flowers spilling from her mouth. The picture reminded the inspector of the gold chain lying in the mouth of one of the Monster's first victims. This clue so captivated Perugini that the cover of the book he would publish about the case showed Botticelli's nymph vomiting bloody flowers.

Perugini organized a twelve-day search of Pacciani's property. The police took apart the farmer's miserable house and plowed up his garden. The haul was pretty disappointing, but on the twelfth day, just as the operation was winding down, Perugini announced with great fanfare that he had found an unfired .22 bullet in the garden. Later, in court, experts said it "might" have been inserted into the infamous Beretta and ejected without being fired—the ballistics report was inconclusive. In Pacciani's garage, the police found a scrap of torn rag, which was duly cataloged. Not long afterward, the Carabinieri received a piece of a .22 Beretta wrapped in a torn rag, with an anonymous note saying it had been found under a tree where Pacciani often went. When the two rags were compared, they matched up.

Pacciani was arrested on January 16, 1993, and charged with being the Monster of Florence. The public, by and large, approved of his arrest. Spezi, however, remained unconvinced. He felt that a drunken, semiliterate peasant given to fits of rage could not possibly have committed the meticulous crimes he had seen. Spezi continued to feel that the Sardinian investigation had been prematurely closed. He laid out his views in a series of carefully reasoned articles, but few readers were persuaded: Pacciani's trial was being broadcast almost every night on television, and the drama of the proceedings overwhelmed all logic. Florentines have never forgotten the sight of Pacciani's violated daughters (one of whom had entered a convent), weeping on the witness stand as they described being raped by their father.

This was melodrama worthy of Puccini. Pacciani rocked and sobbed during the proceedings, sometimes crying out in his Tuscan dialect, "I am a sweet little lamb! . . . I am here like Christ on the cross!" At other times he erupted, face on fire, spittle flying from his lips. Thomas Harris attended the trial, taking notes in longhand on yellow legal pads. The prosecutors presented no murder weapon and no reliable eyewitnesses. Even Pacciani's wife and daughters, who hated him, said he couldn't have been the Monster—he was home drunk most of the time, yelling, hitting them, and acting the bully.

Pacciani was convicted and sentenced to life in prison. During the mandatory appeal, the prosecutor assigned to handle the case did something almost unheard of: he refused to prosecute. He became Pacciani's unlikely advocate, decrying in court the lack of evidence and comparing the police investigation to the work of Inspector Clouseau.

On February 13, 1996, Pacciani was acquitted. A higher court sent the case back to be retried, but Pacciani died in February 1998, before the new trial could begin.

On the very day of Pacciani's acquittal, the police brought forward new witnesses and dramatically announced that they had a confession in hand, in an attempt to salvage their case. The judge refused to allow them to testify, and excoriated the police for the last-minute maneuver. But the investigation was far from over, and eventually their story would emerge. The first surprise witness had actually confessed to being Pacciani's accomplice. He said that he and Pacciani had been hired by a wealthy Florentine doctor to "handle a few little jobs." These "little jobs," investigators later said, were to collect female body parts for Black Masses, to be used as offerings to the devil. The man made many odd claims and contradicted himself continually. He implicated a third man, and the two men were convicted, in a subsequent trial, of murder; one was sentenced to life in prison and the other to twenty-six years.

Thus began the investigation that remains open to this day: the search for the Florentine doctor and the other *mandanti* (masterminds) behind the killings. Inconveniently, the police's self-inculpating witness said he didn't know the doctor's name—he claimed that only Pacciani did. But Pacciani denied the whole story to his dying day.

With the death of Pacciani and the conviction of his accomplices, the investigation receded into the shadows. Most people felt the case had been solved, and Florence moved on. And perhaps it was just as well. For over time, thread by thread, the web of evidence began to unravel. The rag and gun pieces were found to have been a manufactured clue, although by whom was not established. The expert who had been asked to certify that the bullet found in Pacciani's garden might have been inserted into the Monster's gun complained of pressure put on him. On assignment from a television station, Spezi videotaped a police officer, present at the search of Pacciani's property, saying it was his impression that the chief inspector had planted the bullet. The television station refused to air the segment; Spezi published the allegation—and was promptly sued for libel. (He won the case, but not without further antagonizing the Squadra anti-Mostro and its boss.)

Spezi was by now exhausted—by the case, which he had covered for more than fifteen years, and by the grueling life of a crime correspondent. Brother Galileo had urged him to quit his job. Spezi's daughter was growing up, and he was feeling the pinch of his journalist's salary. When his wife's cousin offered him a lucrative partnership in his luggage business, he jumped. *La Nazione* agreed to keep him on a freelance contract. It would leave him time to fulfill a longtime dream of writing mystery novels, and to embark on his own counter-investigation, which became a hobby of sorts.

Around this time, Spezi received crucial help from a high-ranking official in the Carabinieri whose identity he has never revealed, not even to me. This man was part of a group of officers who had continued a secret investigation into the killings after the Carabinieri officially withdrew from the case. His clandestine group had identified a possible suspect as the Monster, a man who had previously been arrested and released.

One of Spezi's big scoops had been the discovery of a report prepared for Inspector Perugini by the FBI's Behavioral Science Unit in Quantico, Virginia. It had been commissioned in secret and then suppressed, because it didn't describe Pacciani. The report cataloged the killer's likely characteristics, explained his probable motive, and speculated as to how and why he killed, how he chose his targets, what he did with the body parts he collected, and much more. Its conclusion was that the Monster was of a type well known to the FBI: a lone, sexually impotent male with a pathological hatred of women, who satisfied his libidinous cravings through killing.

The FBI report said that the Monster chose the *places* for his crimes, not the victims, and that he would kill only in familiar locations. The murders had been committed in some of the most beautiful landscapes in the world, over a large area encompassing the hills south, east, and north of Florence. The police had stared at pins in maps for years, never finding a pattern. But when Spezi mapped the life and movements of the Carabinieri's suspect to the locations of the killings, he found surprising overlaps.

After that first meeting at Caffè Ricchi, Mario Spezi and I became friends. He often talked about the case, and I began to share his frustration at its unsatisfying conclusion. I cannot remember exactly when my curiosity became more than idle speculation, but by the spring of 2001, Spezi and I had agreed to write something together—a collaboration that would eventually take the form of a book. But first, I needed a crash course from the "Monstrologer."

A couple of days a week I would shove my laptop into a backpack and bicycle the six miles from our home to Spezi's. The last kilometer was a bear, almost straight uphill through groves of knotted olive trees. I would find him in the dining room, thick with smoke, with papers and photographs scattered about the table. Myriam, Spezi's wife, would check in on us every now and then and bring us cups of espresso or fresh-squeezed orange juice. Spezi was always careful to keep the strongest details—and the crime-scene photographs—well out of her sight.

He went through the entire history chronologically, chain-smoking all the way, from time to time plucking a document or a photograph from the heap by way of illustration. I took notes furiously on my computer, in an almost indecipherable mix of English and Italian (I was still learning the language then). "*Bello*, eh?" he often said when he had finished recounting some particularly egregious example of investigative incompetence.

We visited the crime scenes together and tracked down, where we could, the family members of the victims. We went to Vicchio to visit the mother of one of the female victims. She was living a hollow existence in what had once been an imposing house in the center of town. Her husband had squandered the family fortune on his futile search for the killer and had dropped dead of a heart attack at police headquarters during one of his many visits.

I began to understand, in a small way, the immensity of the evil that Brother Galileo—now dead—had helped Spezi to accept. And yet, despite the darkness of our hunt, my days with Spezi were my happiest in Italy. My wife and I enjoyed many elegant dinners with Mario and Myriam on their terrace overlooking the hills, where they gathered writers, photographers, countesses—even, one night, a woman who was half Apache, half Florentine. Spezi had a seemingly inexhaustible store of outrageous tales, which he recounted with the quiet delight of an epicure serving a mossy bottle of Château Pétrus. As he told a story, he often imitated the players, speaking in flawless dialect. Sometimes he would require each guest to bring a story to dinner, in lieu of flowers or wine.

Spezi's view of the case was not complicated. He had nothing but contempt for the conspiracy theories and heated speculation about satanic sects. The simplest and most obvious explanation, to his mind, was most likely the correct one. He had always believed—and I came to share his conviction—that the Monster of Florence was a lone psychopath, and that the key to finding him was the gun used in the 1968 clan killing. Every cop, Spezi often told me, knows that a gun used in a homicide—especially a clan killing—is never disposed of casually. It is either destroyed or kept in a safe place. One of the killers had taken the gun home.

Spezi believed that the Monster must either be Salvatore Vinci, the man Rotella had had his eye on, or someone close to him—someone with access to the gun *and* the box of bullets. It was that simple. He turned to the crime-scene evidence. It suggested that the Monster was a tall, right-handed man in excellent physical condition who acted with almost preternatural sangfroid (ruling out Pacciani, who was short, fat, old, and usually drunk). The killer was an expert shot and skilled with a knife.

When we had chased down every other person we could find with some connection to the crimes, I pressed Spezi on the subject of the Carabinieri's suspect, who he told me was the son of one of the original Sardinians. He was still alive and living in Florence. (I will not mention his name, since the evidence against him remains circumstantial.) The stumbling block was Myriam, who had begged her husband not to approach him. Alone, Spezi had heeded her pleas, but there were two of us now, and I goaded him on.

Without telling Myriam, Spezi and I began to plan our visit. We developed a cover story—that I was an American journalist and Spezi my translator, and we were conducting a series of interviews about the Monster of Florence case. Out of deference to Myriam's fears, we decided to use false names.

For years, the Sardinian had been living a quiet life in a working-class area west of Florence. We arrived at his apartment building at 9:40 p.m., when we would be most likely to find him home. His neighborhood was neat, even cheerful, with tiny garden plots in front of modest stuccoed apartment buildings. There was a grocery store on the corner, and bicycles were chained to the railings. Across the street, past a row of umbrella pines, lay the skeletons of abandoned textile factories.

Spezi pressed the intercom, and a woman's voice asked, "Who is it?"

"Marco Tiezzi," Spezi replied.

We were buzzed in without further questions.

A man greeted us at the door, wearing only a pair of tight shorts. He recognized Spezi immediately: "Ah, Spezi! It's you!" he said with a smile. "I must've misheard the name. I've wanted to meet you for the longest time." He invited us to be seated at a small kitchen table and offered us a glass of Mirto, a Sardinian liqueur. His wife, who had been washing spinach in a sink, silently left the room.

Our host was a strikingly handsome man with a dimpled smile. His curly black hair was lightly peppered with gray, his body tanned and heavily muscled. He projected a cocky air of working-class charm. While we talked about the case, he casually rippled the muscles of his upper arms or slid his hands over them in what seemed an unconscious gesture of self-admiration. He spoke in a husky and compelling voice that reminded me of Robert De Niro's.

Spezi casually slipped his tape recorder out of his pocket and laid it on the table. "May I?"

The man flexed and smiled. "No," he said. "I'm jealous of my voice."

Spezi took notes in longhand, slowly working his way from generic questions toward his real objective. (The quoted passages below are from his notes.)

"Your father had strange sexual habits," Spezi said. "Perhaps that was a reason you hated him?"

"Back then I knew nothing about it. Only later did I hear about his . . . tics."

"But you and he had some really big fights. Even when you were young. In the spring of 1974, for example, you were charged with breaking and entering and theft."

"That's not correct. Since he didn't know if I had taken anything, I was charged only with violation of domicile. Another time we had a big fight, and I pinned him, putting my scuba knife to his throat, but he broke free and I locked myself in the bathroom."

"When did you leave Florence?"

"In the beginning of 1975. First I went to Sardinia, and then to Lake Como."

"Then you returned and got married."

"Right. I married a childhood sweetheart, but it didn't work. We married in 1982 and separated in 1985."

"What didn't work?"

"She couldn't have children."

Spezi did not mention he had learned that the marriage had been annulled for non-consummation. "Can I ask you a rather direct question?" he said.

"Sure. I may not answer it."

"If your father owned the .22 Beretta, you were the person in the best position to take it. Perhaps during the breaking and entering in the spring of 1974."

He didn't answer immediately. "I have proof I didn't take it," he said at last.

"Which is?"

"If I had taken it, I would have fired it into my father's forehead."

Spezi pressed on. He pointed out that our host had been away from Florence from 1975 to 1980, when there were no Monster killings. When he returned, they began again.

The Sardinian leaned back in his seat, and his smile broadened. "Those years were the best of my life," he said, "up there at Lake Como. I had a house, I ate well, and all those girls..." He whistled and made a vulgar Italian gesture.

"And so," Spezi said, " you're not... the Monster of Florence."

There was only a brief hesitation. The Sardinian never ceased smiling, not even for an instant. "No," he said. "I like my pussy whole."

We rose to go. Our host followed us to the door. Just before opening it, he leaned toward Spezi and spoke in a low and casual voice. "Ah, Spezi, I was forgetting something." He leaned even closer and smiled. His voice took on a hoarse, gravelly tone. "Listen carefully: I never joke around."

Spezi and I agreed that we would publish our book first in Italian, and then I would rewrite it and publish it in English. The publisher of my murder mysteries in Italy, Sonzogno, a division of the Italian publishing house RCS Libri, gave us a contract and an advance. The book, which was titled *Dolci Colline di Sangue* (*Sweet Hills of Blood*, a play on the phrase *dolci colline di Firenze*), was scheduled for publication in April 2006.

Meanwhile, the search for the hidden masterminds began to intensify. Time had passed, and the old investigators had retired or been promoted. Vigna was appointed head of Italy's anti-Mafia unit, while Perugini went on to become the liaison officer with the American FBI. A new investigator rose to the fore: Chief Inspector Michele Giuttari, who had organized and headed an elite police unit known as GIDES (Investigative Group for Serial Crimes), heir to Perugini's Squadra anti-Mostro. The newspapers dubbed him "*il Superpoliziotto*," because he was, in practice, answerable to nobody.

In the summer of 2001, the case once again hit the front pages in Italy. GIDES had focused its attention on a villa in Chianti where Pietro Pacciani had worked as a gardener. This villa, which the papers dubbed the "Villa of Horrors," was suspected of being the meeting place for the cult of devil worshippers who had supposedly hired Pacciani to do their bidding. One important clue that a satanic sect was behind the killings was a rough, hexagonal stone in the form of a broken pyramid found at the site of one of the crimes. Only Giuttari realized its significance. "I hypothesized that it was an object connected to the occult," he would later explain in one of his many books on the case, "and that, for some reason, it had been left there deliberately."

Spezi ridiculed this conclusion in the media and produced a similar stone, which a friend had given him. He pointed out that it was not an esoteric object at all, but a type of doorstop commonly found in old Tuscan farmhouses. On May 14, he appeared on a popular TV show with an explosive allegation: he had shown the photographs of the murdered French tourists to one of Europe's leading forensic entomologists, and the entomologist had concluded, by examining the larvae on the corpses, that the lovers could not have been killed any later than Saturday night.

This determination, if true, was fatal to the satanic-sect theory: Pacciani's supposed accomplice swore that the French tourists were killed on Sunday night. If the crime had occurred Saturday, all his claims would be thrown into question. What's more, Pacciani had an ironclad alibi for Saturday night.

Much to Spezi's dismay, Giuttari dismissed the entomologist's findings and pressed ahead with his investigation. Spezi's appearance on television had another effect entirely from that intended: it seemed to have inspired the chief inspector's undying hatred.

In June 2004, I moved back to the States with my family, into a house we had built on the coast of Maine. When I left Italy, Mario gave me a pencil drawing he'd made of Pacciani during the trial and a caricature of myself, spying on my wife through a keyhole. I hung both on the wall of my writing shack, in the woods behind our house, along with a photograph of Spezi in his fedora and trench coat, standing in a butcher's shop under a rack of hog jowls.

Spezi and I spoke frequently. I missed my life in Italy—but Maine was quiet, and quiet is what a writer needs. We continued to work on the book by email and telephone. Spezi did most of the actual writing, while I made suggestions and contributed a few chapters, which he had to rewrite (I write in Italian at about a fifth-grade level). He continued to keep me abreast of the *"Pista Satanica,"* which, curiously, seemed to be heating up.

That summer, Spezi called me with a strange bit of news. An old friend of his, a pharmacist, was being investigated for the death of Francesco Narducci, a gastroenterologist whose body had been found floating in Lake Trasimeno some twenty years earlier. The original investigators had considered it a suicide—Narducci had been heavily into drugs and was known to be depressed—but GIDES suspected that the doctor may in fact have been murdered by the Monster's satanic sect. This brought into the investigation the public prosecutor who has jurisdiction over Lake Trasimeno: the *pubblico ministero* of Perugia, Giuliano Mignini.

On November 18, 2004, at six a.m., Spezi and his family were awoken by the sound of their door buzzer. *"Polizia!"* screamed a voice. *"Perquisizione!"* The police were from GIDES, Giuttari's squad. Their warrant gave two reasons for the search: Spezi had "materially damaged the investigation by casting doubt on the accusations through use of the medium of television," and he had "evidenced a peculiar and suspicious interest in . . . the investigation." He was served with an *avviso di garanzia*, one step short of a formal indictment. It listed seventeen crimes for which Spezi was being investigated, all undisclosed.

For seven hours the police searched the apartment, while Spezi, Myriam, and their daughter looked on. Officers pulled books off the shelves and rummaged through photos, letters, and schoolbooks, scattering things on the floor. They took everything Spezi had that related to the case: his computer, disks, archives, clippings, interviews, even our notes and drafts of the book. They found Spezi's old doorstop, which a document would later describe as having been "secreted behind a door." To Giuttari, this became one of the most important fruits of the search.

Twelve months later, Spezi opened the newspaper and read a headline about himself: "Narducci Murder: Journalist Investigated."

"When I read that," Spezi told me, "it was like a hallucination. I felt I was inside a film of Kafka's *Trial*, remade by Jerry Lewis and Dean Martin." Spezi had gone from journalist to suspect.

It was in this climate that I arrived in Florence on February 14, 2006. The kids were on winter vacation, and plane tickets were cheap. I was anxious to see my old friend. Our book would be published in two months, and Spezi hoped we might do some preliminary publicity and line up an evening presentation at Seeber, one of the best bookstores in Florence.

I visited Spezi on February 15. He told me he had recently heard from a source that during the Monster killings and afterward, a group of Sardinians had frequented a run-down outbuilding on the thousand-acre estate of a grand villa outside Florence. The source claimed to have a friend who had been at the building a few months earlier, with the Carabinieri's suspect (the man I've been calling "the Sardinian"). He had seen six locked iron boxes and two guns: a machine pistol and a .22 Beretta.

"What are those boxes?" the friend had asked.

"That's *my* stuff," the Sardinian allegedly said, slapping his chest.

Six locked iron boxes. Six female victims. A .22 Beretta. It was almost too perfect to be true.

I asked Spezi what he planned to do. He said he had been agonizing about this. He smelled a scoop—the scoop of a lifetime. He had driven past the villa a couple of times, but in the end he had decided that he had no choice but to call the police.

I had never seen Mario so excited. "This could be it," he told me. "The culmination of all my years on this case. And you'll be here to see it." He said the villa was open to the public for sales of wine and olive oil, and that it was rented out for parties.

I asked if I could see it. "Sure," he said. "Why not?" We couldn't go to the outbuilding, but we could at least see the part open to the public.

We decided to go the next day, accompanied by a friend of Spezi's who owned a security firm in Florence and from his days as a cop knew the source, an ex-con. Driving over to his friend's office, Spezi apologized for the state of his car: a few days earlier someone had wrecked the door and stolen his radio.

It was raining when we arrived at the villa. A woman leaned out a window and said the salesroom was closed for lunch. We took a desultory stroll along the drive and returned to the car. We had been there about ten minutes. It was a disappointing visit, at least to me; something about the whole story didn't feel quite right.

Two days later, Spezi called me on my cell phone. "We did it," he said. "We did it all." He didn't go into details, but I knew what he meant: he had given the information to the police. He also said, before I could ask too many questions, "The telephone is bad." For two years he had been complaining that the police were tapping his phones.

On February 22, as I was heading out for a morning coffee, my cell phone rang. A man speaking Italian informed me that he was a police detective and that he needed to see me—immediately. No, it wasn't a joke. And no, he couldn't tell me what it was about, only that it was "*obbligatorio*."

I chose the most public place possible, the Piazza della Signoria. Two plainclothes detectives from GIDES took me into the Palazzo Vecchio, where, in the magnificent Renaissance courtyard, surrounded by Vasari's frescoes, I was presented with a legal summons to appear before Judge Mignini. The detective politely explained that a no-show would be a serious crime; it would put him in the regrettable position of having to come and get me.

I asked, "Is this about the Monster of Florence case?"

"*Bravo*," said the detective.

The next day, I was ushered into a pleasant office in the Procura della Repubblica, just outside the ancient city walls of Perugia. Present were one of the detectives from the previous day, a small and very tense captain of police with orange hair, a stenographer, and Giuliano Mignini, sitting behind a desk. I had dressed smartly—Italians judge harshly in such matters—and I had a folded copy of the *International Herald Tribune* under my arm as a prop.

Mignini was a small man of indeterminate middle age, well groomed, with a fleshy face and thinning hair. His voice was calm and pleasant and he addressed me with elaborate courtesy, bestowing the honorific of *dottore*, which in Italy denotes the highest respect. He explained that I had the right to an interpreter, but finding one might take many hours, during which time I would be unpleasantly detained. In his opinion, I spoke Italian fluently. I asked if I needed a lawyer, and he said that, although it was of course my right, it wasn't necessary; he merely wanted to ask a few questions of a routine nature.

His questions were gentle, posed almost apologetically. The stenographer typed the questions, and my answers, into her computer. Sometimes Mignini rephrased my answers in better Italian, checking solicitously to see if that was what I had meant to say. He asked me about Spezi's lawyer, Alessandro Traversi, and wanted to know what I could say about Spezi's legal strategy. He named many names and asked if Spezi had ever mentioned them. Most were unfamiliar. The questions went on like this for an hour, and I was starting to feel reassured. I even had a glimmer of hope that I might get out in time to join my wife and children for lunch at a nearby restaurant, which came highly recommended in the guidebooks.

At this point the conversation turned to our visit to the villa. Why did we go? What did we do there? Where exactly did we walk? Was there talk of a gun? Of iron boxes? Was my back ever to Spezi? Did we see anyone there? Who? What was said?

I answered truthfully, trying to suppress a damnable habit of over-explanation, but I could see that Mignini was not happy. He repeated the same questions, in different forms. It began to dawn on me that the previous line of inquiry had been nothing more than a few balls lobbed in the bullpen. Now the game had begun.

Mignini's face flushed as his frustration mounted. He frequently instructed the stenographer to read back my earlier answers. "You said that, and now you say this. Which is true, Dottor Preston? *Which is true?*"

I began to stumble over my words (as I've noted, I am not fluent in Italian, especially legal and criminological terms). With a growing sense of dismay, I could hear from my own stammering, hesitant voice that I was sounding like a liar.

"Listen to this," Mignini said. He nodded to the stenographer, who pressed a button on her computer. There was the ringing of a phone, and then my voice:

"*Pronto.*"

"*Ciao, sono Mario.*"

Spezi and I chatted for a moment while I listened in amazement to my own voice, clearer on the intercept than in the original call on my lousy cell phone. Mignini played it once, then again. He stopped at the point where Spezi said, "We did it all," and fixed his eyes on me: "What exactly did you do, Dottor Preston?"

I explained that Spezi was referring to his decision to report to the police what he had heard about possible evidence hidden at the villa.

"No, Dottor Preston." He played the recording again and again, asking repeatedly, "What is this thing you did? *What did you do?*" He seized on Spezi's comment that the telephone was bad. What did he mean by that?

I explained that he thought the phone was tapped.

And why, Mignini wanted to know, were we concerned about the phones being tapped *if we weren't engaged in illegal activity?*

"Because it isn't nice to have your phone tapped," I answered feebly.

"That is not an answer, Dottor Preston."

He played the recording again, stopping at several words and demanding to know what Spezi or I meant, as if we were speaking in code, a common Mafia ploy. I tried to explain that the conversation meant what it said, but Mignini brushed my explanations aside. His face was flushed with a look of contempt. I knew why: he had expected me to lie, and I had met his expectation. I stammered out a question: Did he think we had committed a crime at the villa?

Mignini straightened up in his chair and, with a note of triumph in his voice, said, "*Yes*."

"What?"

"You and Spezi either planted, or were planning to plant, *false evidence* at that villa in an attempt to frame an innocent man for being the Monster of Florence, to derail this investigation, and to deflect suspicion from Spezi. *That* is what you were doing. This comment—*We did it all*—that is what he meant."

I was floored. I stammered that this was just a theory, but Mignini interrupted me and said, "These are not theories. They are facts!" He insisted I knew perfectly well that Spezi was being investigated for the murder of Narducci, and that I knew more about the murder than I was letting on. "That makes you an accessory. Yes, Dottor Preston," Mignini insisted, "I can *hear* it in your voice. I can hear the tone of knowledge, of deep familiarity with these events. Just listen." His voice rose with restrained exaltation. "*Listen* to yourself!"

And, for maybe the tenth time, he replayed the phone conversation. "Perhaps you have been duped, but I don't think so. You *know*! And now, you have one last chance—one *last* chance—to tell us what you know, or I will charge you with perjury. I don't care; I will do it, even if the news goes around the world tomorrow."

I felt sick, and I had the sudden urge to relieve myself. I asked for the way to the bathroom. I returned a few minutes later, having failed to muster much composure. "I've told you the truth," I managed to croak. "What more can I say?"

Mignini waved his hand and was handed a legal tome. He placed it on his desk with the utmost delicacy, opened it, and, in a voice worthy of a funeral oration, began to read the text of the law. I heard that I was now "*indagato*" (an official suspect under investigation) for the crime of reticence and making false statements. He announced that the investigation would be suspended to allow me to leave Italy, but that it would be reinstated when the investigation of Spezi was concluded.

The secretary printed out a transcript. The two-and-a-half-hour interrogation had been edited down to two pages, which I amended and signed.

"May I keep this?" I asked.

"No. It is under seal."

Very stiffly, I picked up my *International Herald Tribune*, folded it under my arm, and turned to leave.

"If you ever decide to talk, Dottor Preston, we are here."

On rubbery legs I descended to the street, into a wintry drizzle.

I left Italy the next day. When I returned to my home in Maine, which stands on a bluff overlooking the gray Atlantic, and listened to the breakers on the rocks below and the seagulls calling above, I felt tears trickling down my face.

But it was not over—not at all.

After I left, Spezi brought his car to a mechanic to get the broken door and radio fixed. The mechanic emerged holding a few thousand dollars' worth of electronics: a sophisticated GPS, microphone, and transmitter, which had been carefully attached to the old radio wires. Spezi filed a complaint, and a week or two later his crappy radio was returned to him by GIDES.

For Spezi, the wrecking of his car was the last straw. He asked his lawyer to file a civil lawsuit against Chief Inspector Michele Giuttari. The suit was dated March 23. Spezi wrote the introductory statement himself, every word perfectly pitched to infuriate his foe:

> For more than a year, I have been the victim not just of half-baked police work, but of what could be said to be authentic violations of civil rights. This phenomenon—which pertains not just to me, but to many others—brings to mind the most dysfunctional societies, such as one might expect to find in Asia or Africa.

Spezi proceeded to deliver an uppercut to Giuttari's soft underbelly—his literary talent. In February, Giuttari had published his second book (there had also been several novels) on the Monster of Florence case, *The Monster: Anatomy of an Investigation*, in which he had taken several jabs at Spezi and others. In the lawsuit, Spezi quoted extract after extract, savaging Giuttari's theories, his logic, and his writing ability.

On Friday, April 7, eleven days before the publication of our book, a squad of policemen arrived at Spezi's apartment, lured him outside under false pretenses, arrested him, and hustled him into a car. He was driven to GIDES headquarters and taken from there to prison in Perugia. The Italian papers reported the charges against him: "calumny," "disturbing an essential public service," and "attempting to derail the investigation into the case of the Monster of Florence." A number of other people were named by the police as being involved in these crimes; I was one of them. The final charge, the papers claimed, was complicity in murder.

The day of the arrest, Mignini asked for and received a special dispensation to invoke a law that is normally used only for terrorists or Mafia dons who pose an imminent threat to the state. For a period of five days Spezi was denied access to his lawyers, kept in a tiny isolation cell under conditions of extreme deprivation, and grilled mercilessly. It was noted in the press that Spezi's treatment was harsher than that of Bernardo Provenzano, the Mafia "boss of bosses" captured in Sicily a few days later.

Spezi spent three weeks in Capanne, one of Italy's grimmest prisons. On April 29, a three-judge panel in Perugia surprised everyone by annulling his imprisonment and setting him free. It was a decisive slap in the face for Mignini and Giuttari. A week later, Florence hosted a demonstration for freedom of the press, and Spezi was the guest of honor. That same day our book hit the bestseller lists across Italy.

When Spezi returned home from prison, a crowd of journalists greeted him. "No, I'll not deal with the Monster affair anymore," he told one. "I'll write books, but not about that." Twenty-five years after that perfect summer Sunday in June when the bodies of two lovers were first found, Mario Spezi had finally declared his emancipation from the Monster of Florence.

Spezi's legal problems will likely drag on for years. He has been summoned back for another round of interrogation, and fresh charges are reportedly in the works. And yet, the tide may be turning. Mignini's fellow judges have severely criticized his conduct of the case, and, in early May, Giuttari himself became the target of an investigation, accused of falsifying evidence in the case. And so the investigation grinds on, voracious in its appetite for new victims.

People have often asked me if the Monster of Florence will ever be found. I once believed that Spezi and I could find the truth; now I am not so certain. Any crime novel, to be successful, must contain certain basic elements: there must be a motive; evidence; a trail of clues; and a process of discovery that leads, one way or another, to a conclusion. All novels, even *Crime and Punishment*, must come to an end.

But life, I have learned, is not so tidy. Here were murders without motive and a trail of clues apparently without end. The process of discovery has led investigators so deeply into a wilderness of falsehood that I doubt they will ever find their way out. Spezi and I used to laugh at their elaborate theories, but ours may not be much better. It wasn't based on what a good criminal investigation should be: the nitty-gritty of blood, hair, fibers, fingerprints, DNA, and reliable eyewitnesses. In the absence of solid forensic police work—which, in the Monster case, was shockingly deficient—any hypothesis will remain like something dreamed up by Hercule Poirot: a beautiful story in search of a confession. Only this is not a novel, and there won't be a confession—and without one, the Monster of Florence will never be found.

UPDATE

This story fascinated me in large part because it promised to shed light on the mystery of human evil and how it can exist. In the years since I wrote this article and the book that followed, I finally came to believe that human evil is an absence instead of a presence—a subtraction, not an addition. Half a century ago, the philosopher Hannah Arendt wrote famously about the "banality of evil" in reporting on the war crimes of Adolf Eichmann. She called him "terrifyingly normal" and "shallow," saying he lacked the ability "to think from the standpoint of somebody else." The same, I think, can be said about the evil of the Monster. His profound depravity was due to an absence or subtraction of human attributes from his being—conscience, compassion, and empathy—not the addition of something twisted. He was a partial human being, a ghost, a human residuum instead of a complete person.

Only weeks after the Monster of Florence article was published in The Atlantic, *Italian journalists hunted down and identified the unnamed Sardinian in our story as Antonio Vinci, son of Salvatore Vinci. Mario and I then wrote a nonfiction book on the case, called* The Monster of Florence, which was published in 2008. The book spent four months on the New York Times bestseller list and won many awards. It is currently being made into a television series.

Giuliano Mignini was indicted for abuse of office in the Monster case, which led, indirectly, to his involvement in another sensational murder case, that of an English student in Perugia named Meredith Kercher. (See "Trial by Fury" on page 197.) Police suspicion quickly focused on her roommate, Amanda Knox, a young American student studying in Perugia, and Amanda's Italian boyfriend, Raffaele Sollecito. Amanda was hauled into the police station and interrogated for over fifteen hours in Italian, a language she hardly knew, smacked on the head, and finally coerced to sign a false quasi-confession, written out in police-jargon Italian which she couldn't possibly have been able to read. She and her boyfriend and a third person were arrested for the murder. Mignini and the Perugian police chief then made a triumphant announcement: caso chiuso—*case closed*—and the news went around the world.

Mignini, in his eagerness to redeem himself from the indictment in the Monster case, had made a fatal mistake in declaring the case solved before the crime scene had been forensically processed. When the results came in, investigators found no DNA anywhere at the scene from Amanda or Raffaele. Instead, they found DNA and other damning evidence of an unknown person all over the room—his handprint in her blood, his DNA in her purse from which money had been stolen, and his DNA inside of her, as she had been sexually assaulted. The DNA belonged to a drifter and petty criminal from the Ivory Coast who lived in Perugia named Rudy Guede. Guede, who had fled Italy the day after the murder, was arrested and brought back from Germany.

A colossal injustice followed. Rather than release Amanda and Raffaele and admit the mistake, to save his career, already teetering from the abuse of office indictment in the Monster case, Mignini proceeded to assert that Guede had committed the crime *with* Amanda and Raffaele, claiming that Meredith was murdered by them in a satanic ritual. (He later retracted that claim after a torrent of ridicule.) Mignini, Perugian police officers, and flexible forensic labs then proceeded to frame Amanda and Raffaele for the murder using manipulated evidence, false DNA results, and highly questionable testimony, while suppressing all exculpatory facts. After spending four years in prison, Amanda and Raffaele were finally acquitted by the Italian Supreme Court, which severely criticized the prosecutors and wrote the case was characterized by "sensational investigative failures." In this way, the Monster of Florence case and the Amanda Knox case are linked.

After Mario Spezi was released from prison and the charges dropped, Mignini repeatedly brought fresh charges against him, all of which were dismissed by the courts, but which had the effect of draining Mario's finances and preventing him from working as a journalist. Mario died on September 9, 2016, after a long illness, which his friends and relatives believe was aggravated by Mignini's vindictive serial persecution of him.

Antonio Banderas has been cast as Mario Spezi in the television series, directed by Nikolaj Arcel, which is expected to air in 2024.

UNEXPLAINED DEATHS

THE SKELETONS AT THE LAKE

IN THE WINTER of 1942, on the shores of a lake high in the Himalayas, a forest ranger came across hundreds of bones and skulls, some with flesh still on them. When the snow and ice melted that summer, many more were visible through the clear water, lying on the bottom. The lake, a glacial tarn called Roopkund, was more than sixteen thousand feet above sea level, an arduous five-day trek from human habitation, in a mountain cirque surrounded by snowfields and battered by storms. In the midst of the Second World War, British officials in India initially worried that the dead might be the remains of Japanese soldiers attempting a secret invasion. The apparent age of the bones quickly dispelled that idea. But what had happened to all these people? Why were they in the mountains, and when and how had they died?

In 1956, the Anthropological Survey of India, in Calcutta, sponsored several expeditions to Roopkund to investigate. A snowstorm forced the first expedition to turn back, but two months later another expedition made it and returned to Calcutta with remains for study. Carbon dating, still an unreliable innovation, indicated that the bones were between five hundred and eight hundred years old.

Indian scientists were intensely interested in the Roopkund mystery. The lake, some thought, was a place where holy men committed ritual suicide. Or maybe the dead were a detachment of soldiers from

---

Originally published in the *New Yorker* in 2020.

a thirteenth-century army sent by the Sultan of Delhi in an ill-fated attempt to invade Tibet, or a group of Tibet-bound traders who had lost their way. Perhaps this was hallowed ground, an open-air cemetery, or a place where victims of an epidemic were dumped to prevent contagion.

People in the villages below Roopkund had their own explanation, passed down in folk songs and stories. The villages are on the route of a pilgrimage to honor Nanda Devi, a manifestation of Parvati, a supreme goddess in Hinduism. The pilgrimage winds up through the foothills of the Trisul massif, where locals believe that the goddess lives with her husband, Shiva. It may be the longest and most dangerous pilgrimage in India, and a particularly perilous section—the Jyumra Gali, or Path of Death—runs along a ridge high above Roopkund. As the villagers tell it, long ago Nanda Devi left her home to visit a distant kingdom, where she was treated discourteously by the king and queen. Nanda Devi cursed the kingdom, unleashing drought and disaster, and infesting the milk and rice with maggots. In order to appease the goddess, the royal couple embarked on a pilgrimage. The king, who liked his entertainments, took along a bevy of dancing courtesans and musicians, in violation of the ascetic traditions of the pilgrimage. Nanda Devi was furious at the display of earthly pleasures, and she shoved the dancing girls down into the underworld. The pits into which they are said to have sunk are still visible high on a mountainside. Then, according to the legend, she sent down a blizzard of hail and a whirlwind, which swept all the pilgrims on the Path of Death into the lake. Their skeletons are a warning to those who would disrespect the goddess.

This story is retold in *Mountain Goddess*, a 1991 book by the American anthropologist William Sax. Now a professor at Heidelberg University, he stumbled upon a reference to the lake and the bodies as an undergraduate, in the 1970s, and was fascinated. He and a friend traveled to the hamlet of Wan, the settlement closest to Roopkund, where a local man agreed to guide them up the pilgrim trail to the lake. The trail climbs through deep forests, emerging above the tree line, at eleven and a half thousand feet, into meadows carpeted with wildflowers. To the north is a vast wall of Himalayan peaks, some of the highest in the world. From there, the route follows steep ridgelines and leads past an ancient stone shrine, festooned with bronze bells and tridents and containing a statue of the elephant deity Ganesha. Then, at fifteen thousand feet, it goes over a pass and up a series of switchbacks through scree to Roopkund. The lake, about a hundred and thirty feet across and ten feet deep, is an emerald jewel nestled in a bowl of rock and ice. (In Hindi, *roop kund* means "beautifully shaped lake.") Almost as soon as Sax and his companions arrived, they were engulfed by a blizzard and stumbled around the bone-strewn cirque in whiteout conditions, calling for one another and nearly adding their own bodies to the charnel ground.

Exhausted and feverish, Sax barely made it back to Wan with his companions, and spent ten days recovering in his guide's stone hut. Yet his passion for the place was undimmed. He went on to write a doctoral thesis about the local traditions surrounding Nanda Devi. In the late eighties, he went on the pilgrimage himself, the only Westerner to have done so at that time, after which he published *Mountain Goddess*. The book describes how the Himalayas, "associated for thousands of years in India's literatures with famous pilgrimage places and powerful, ascetic renouncers," became the setting for followers to show devotion to the goddess by "giving suffering" to their bodies.

In 2005, Sax was featured in a National Geographic documentary about the lake. The Indian media company that made the film assembled a team of archaeologists, anthropologists, geneticists, and technicians from research laboratories in India and the U.K. to collect and study the bones. In the decades since Sax first visited, the lake had become a popular destination in the trekking community and the site was being ruined. Bones had been stolen; others had been rearranged in fanciful patterns or piled in cairns. Almost none of the skeletons were intact, and it was impossible to tell which bones belonged together or where they had originally lain. Nature had added to the confusion, churning and fracturing the bones with rock slides and avalanches. But a recent landslide had exposed a cache of fresh bones and artifacts. Under a slab of rock, the team found the remains of a woman, bent double. The body was intact and still had skin and flesh. The scientists removed tissue samples for testing, shot video, and collected bones and artifacts. The team estimated that the area contained the remains of between three hundred and seven hundred people.

The scientific analysis swiftly discounted most of the prevailing theories. These were not the remains of a lost army: the bones were from men, women, and children. Aside from a single iron spearhead, no weapons were found, and there was no trace of horses. The bones showed no evidence of battle, ritual suicide, murder, or epidemic disease. Nor was Roopkund a cemetery: most of the individuals were healthy and between eighteen and thirty-five years old. Meanwhile, the team's geographic analysis laid to rest the idea of traders lost in the mountains, establishing that no trade route between India and Tibet had ever existed in the area. Although the Tibetan border is only thirty-five miles north of Roopkund, the mountains form an impassable barrier. Besides, no trade goods or beasts of burden were found with the bodies. Artifacts retrieved included dozens of leather slippers, pieces of parasols made of bamboo and birch bark, and bangles made of seashells and glass. Devotees of Nanda Devi carry parasols and wear bangles on the pilgrimage. The dead, it appeared, were most likely pilgrims.

DNA analysis showed that all the victims appeared to have a genetic makeup typical of South Asian origin. Bone and tissue samples were sent to Oxford University for carbon dating. The new dates, far more accurate than the 1956 ones, formed a tight cluster in the ninth century. Tom Higham, who performed the analysis, concluded that the victims had perished in a single event and had "died instantaneously within hours of one another." Meanwhile, a team of bioarchaeologists and paleopathologists noted the presence of two distinct groups: there were "rugged, tall" people with long heads and also some "medium height, lightly built, round headed" people, who displayed a curious shallow groove across the vault of the skull. The scientists concluded that the dead represented two populations: a group of tall Brahmans from the plains of India and a company of shorter, local porters, whose skulls were marked by years of carrying heavy loads with a tumpline looped over their heads.

The investigation also revealed that three or possibly four skulls had compression fractures on the crown that had probably occurred at the time of death. "It is not a weapon injury," the researchers noted, but came "from a blow from a blunt and round heavy object." This stretch of the Himalayas is notorious for hailstorms, which destroy crops and damage property. The team concluded that, around the year 800 A.D., a group of pilgrims were caught in a storm on the exposed ridge above Roopkund and were pummeled to death by giant hailstones. Over the years, landslides and avalanches had rolled the bodies down the steep slope into the lake and the surrounding area. Not only did the mystery of Roopkund appear to be solved; it also seemed that the local tales of Nanda Devi's wrath had originated in an actual event.

In 2019, however, *Nature Communications* published the baffling results of a new study conducted by sixteen research institutions across three continents. Genetic analysis and new carbon dating revealed that a significant proportion of the Roopkund remains belonged to people from somewhere in the eastern Mediterranean, most likely near Crete, and that they had perished at the lake only a couple of centuries ago.

INDIA IS AN ideal country for studying human genetics, ancient and modern. There are fewer cultural barriers to handling human biological materials than in many parts of the world, and Indian scientists have eagerly pursued research into the peopling of the subcontinent. Geneticists have sampled the DNA of hundreds of living populations, making India one of the most genetically mapped countries in the world. In 2008, David Reich, a geneticist at Harvard, made the first of many trips to the country, and visited a leading life-science research institution, the Centre for Cellular and Molecular Biology, in Hyderabad. While there, he discussed someday collaborating on a more detailed study of the Roopkund bones with the center's director, Lalji Singh, and Kumarasamy Thangaraj, a geneticist who had headed up the previous DNA analysis. By the time work began, in 2015, the team, led by the Reich lab and the laboratory in Hyderabad, also included researchers at Pennsylvania State University, the Broad Institute of MIT and Harvard, the Max Planck Institute for the Science of Human History, and the Anthropological Survey of India, where many of the Roopkund bones reside.

Not long before the Covid-19 pandemic shut down the U.S., I visited Reich at Harvard Medical School. His office is a minimalist space with a whiteboard, a table, and a wall of glass looking across Avenue Louis Pasteur to the red brick façade of the Boston Latin School. Reich is a lean, fit man in his mid-forties who speaks with rapid, quiet precision. His self-deprecating manner conceals a supremely self-confident iconoclast who is not averse to toppling received wisdom, and his work has attracted criticism from some anthropologists, archaeologists, and social scientists. The Reich lab, the foremost unit in the country for research into ancient DNA, is responsible for more than half the world's published data in the field. Having so far sequenced the DNA of more than ten thousand long-dead individuals from all over the globe, the lab is almost halfway through a five-year project to create an atlas of human migration and diversity, allowing us to peer deep into our past. The work has produced startling insights into who we are as a species, where we have come from, and what we have done to one another. Hidden in the human genome is evidence of inequality, the displacement of peoples, invasion, mass rape, and large-scale killing. Under the scrutiny of science, the dead are becoming eloquent.

In 2019, Reich led a team of more than a hundred researchers who published a study in *Science* that examined the genomes of some 270 ancient skeletons from the Iberian Peninsula. It's long been known that, from around 2500 to 2000 B.C., major new artistic and cultural styles flourished in western and central Europe. Archaeologists have tended to explain this development as the result of cultural diffusion: people adopted innovations in pottery, metalworking, and weaponry from their geographic neighbors, along with new burial customs and religious beliefs. But the DNA of Iberian skeletons dating from this period of transformation told a different story, revealing what Reich describes as the "genetic scar" of a foreign invasion.

In Iberia during this time, the local type of Y chromosome was replaced by an entirely different type. Given that the Y chromosome, found only in males, is passed down from father to son, this means that the local male line in Iberia was essentially extinguished. It is likely that the newcomers perpetrated a large-scale killing of local men, boys, and possibly male infants. Any local males remaining must have been subjugated in a way that prevented them from fathering children, or were so strongly disfavored in mate selection over time that their genetic contribution was nullified. The full genetic sequencing, however, indicated that about sixty percent of the lineage of the local population was passed on, which shows that women were not killed but almost certainly subjected to widespread sexual coercion, and perhaps even mass rape.

We can get a sense of this reign of terror by thinking about what took place when the descendants of those ancient Iberians sailed to the New World, events for which we have ample historical records. The Spanish conquest of the Americas produced human suffering on a grotesque scale—war, mass murder, rape, slavery, genocide, starvation, and pandemic disease. Genetically, as Reich noted, the outcome was very similar: in Central and South America, large amounts of European DNA mixed into the local population, almost all of it coming from European males. The same Y-chromosome turnover is also found in Americans of African descent. On average, a Black person in America has an ancestry that is around eighty percent African and twenty percent European. But about eighty percent of that European ancestry is inherited from white males—genetic testimony to the widespread rape and sexual coercion of female slaves by slaveowners.

In the Iberian study, the predominant Y chromosome seems to have originated with a group called the Yamnaya, who arose about five thousand years ago in the steppes north of the Black Sea and the Caspian Sea. By adopting the wheel and the horse, they became powerful and fearsome nomads, expanding westward into Europe as well as east- and southward into India. They spoke proto-Indo-European languages, from which most of the languages of Europe and many South Asian languages now spring. Archaeologists have long known about the spread of the Yamnaya, but almost nothing in the archaeological record showed the brutality of their takeover. "This is an example of the power of ancient DNA to reveal *cultural* events," Reich told me.

It also shows how DNA evidence can upset established archaeological theories and bring rejected ones back into contention. The idea that Indo-European languages emanated from the Yamnaya homeland was established in 1956, by the Lithuanian-American archaeologist Marija Gimbutas. Her view, known as the Kurgan hypothesis—named for the distinctive burial mounds that spread west across Europe—is now the most widely accepted theory about Indo-European linguistic origins. But, where many archaeologists envisaged a gradual process of cultural diffusion, Gimbutas saw "continuous waves of expansion or raids." As her career progressed, her ideas became more controversial. In Europe previously, Gimbutas hypothesized, men and women held relatively equal places in a peaceful, female-centered, goddess-worshipping society—as evidenced by the famous fertility figurines of the time. She believed that the nomads from the Caspian steppes imposed a male-dominated warrior culture of violence, sexual inequality, and social stratification, in which women were subservient to men and a small number of elite males accumulated most of the wealth and power.

The DNA from the Iberian skeletons can't tell us what kind of culture the Yamnaya replaced, but it does much to corroborate Gimbutas's sense that the descendants of the Yamnaya caused much greater disruption than other archaeologists believed. Even today, the Y chromosomes of almost all men of western European ancestry have a high percentage of Yamnaya-derived genes, suggesting that violent conquest may have been widespread.

THE TEAM MEMBERS of the Roopkund study planned a variety of tests for the bones. DNA sequencing would show the ancestry of the victims and whether they were related to one another, and carbon dating would estimate when they died. The researchers would test for disease, and analyze the chemistry of the bones to determine the victims' diet and where they might have grown up. Under sterile conditions, the scientists in Hyderabad drilled into long bones and teeth, producing a powder. Vials of this were sent to Harvard and to other labs in India, the United States, and Germany.

An ancient human bone is packed with DNA, but, in many cases, ninety-nine percent or more of that is not human. It is the DNA of billions of microbes that colonized the body during the decomposition after death. To tease the tiny fraction of human DNA from this mass of microbial debris requires a chemical ballet of enormous delicacy, and the risk of contamination is high. Stray DNA molecules from people who handled the remains can ruin an entire sample.

David Reich's lab has a "clean room" for extracting and processing DNA from human tissue. Personnel pass through a dressing area, where they don a full-body clean suit with booties and hood, double pairs of nitrile gloves (the inner one sealed to the suit with tape around the wrists), a hairnet, a face mask, and a plastic shield. The clean room is maintained at positive pressure, which keeps the airflow directed outward, to curtail the entry of airborne DNA. After anything is touched in the room, the outer pair of gloves must be stripped off and a fresh pair put on, in order to prevent the transfer of DNA from surface to surface. Intense ultraviolet light shines whenever the room is empty, to destroy stray DNA. The light is shut off when the lab is occupied, because it burns human skin and eyes.

When I visited, a technician was working on a nubbin of bone from an ancient Roman who lived in Belgium. The whine of a sandblaster filled the air as she removed excess bone from a tiny treasure chest of DNA—a spiral cavity in the inner ear called the cochlea. The bone in which the cochlea is embedded is the densest in the body, and provides the best source of preserved DNA in ancient remains. DNA this old breaks up into short strands. Getting enough to sequence requires complex processes, one of which involves placing samples in a machine that produces a polymerase chain reaction, copying the fragments up to a billion times. The lab doesn't sequence the entire DNA molecule, much of which is repetitious and uninformative, but maps about a million key locations.

Reich had asked a graduate student in his lab, Éadaoin Harney, to take charge of the Roopkund project. Her role was to analyze the Roopkund DNA, wrangle the worldwide team, assemble the results, and write the resulting paper as its lead author. (She has since taken a job as a postdoctoral researcher at the genomics firm 23andMe.) By the middle of 2017, it was apparent that the Roopkund bones belonged to three distinct groups of people. Roopkund A had ancestry typical of South Asians. They were unrelated to one another and genetically diverse, apparently coming from various areas and groups in India. Roopkund C was a lone individual whose genome was typical of Southeast Asia. It was the Roopkund B group, a mixture of men and women unrelated to one another, that confounded everyone. Their genomes did not look Indian or even Asian. "Of all places in the world, India is one of the places most heavily sampled in terms of human diversity," Reich told me. "We have sampled three hundred different groups in India, and there's nothing there even close to Roopkund B."

Harney and Reich began exploring the ancestry of the Roopkund B group, comparing the genomes with hundreds of present-day populations across Europe, Asia, and Africa. The closest match was with people from the Greek island of Crete. "It would be a mistake to say these people were *specifically* from Crete," Reich said. "A very careful analysis showed they don't match perfectly. They are clearly a population of the Aegean area." The Roopkund B group made up more than a third of the samples tested—fourteen individuals out of thirty-eight. Since the bones at the lake were not collected systematically, the finding hinted that the Mediterranean group in total might have been quite large. One-third of three hundred, the lower estimate of the Roopkund dead, is a hundred people.

As bizarre as the result seemed, it nonetheless matched an analysis of bone collagen that the Max Planck Institute and the Harvard lab had done on the same individuals, to determine their diet. Dietary information is stored in our bones, and plants, depending on how they fix carbon during photosynthesis, create one of two chemical signatures—C3 or C4. A person who eats a diet of C3 plants, such as wheat, barley, and rice, will have isotope ratios of carbon in their bones different from those of a person eating a diet high in millets, which are C4. Sure enough, the analysis of Roopkund bone collagen revealed that, in the last ten or so years of their lives, the Roopkund A people ate a varied C3 and C4 diet, typical for much of India; Roopkund B ate a mostly C3 diet, typical of the Mediterranean.

During the study, the Reich lab had divided up its bone-powder samples, sending one portion to the carbon-14-dating laboratory at Penn State. (Doing this rather than having the Penn State samples sent straight from Hyderabad was a way of insuring that the labs were working on the same individuals.) When the carbon-dating results came back, there was another surprise: there appeared to have been multiple mass-death events at Roopkund. The Roopkund A individuals probably died in three or possibly four incidents between 700 and 950 A.D. The Roopkund B group—from the Mediterranean—likely perished in a single event a thousand years later. Because carbon-14 dating is difficult to interpret for the period between 1650 and 1950, the deaths could have occurred anytime during that span, but with a slightly higher probability in the eighteenth century. The lone person of Southeast Asian ancestry in Roopkund C died around the same time.

The eighteenth-century date was so unexpected that Reich and Harney at first thought it might be a typo, or that the samples had been contaminated. Harney wrote up the findings, in a paper co-authored by twenty-seven other scientists. She told me, "We hoped that after the paper was published someone would come forward with information that would help us determine what might have happened at Roopkund—some historian or a person with knowledge of a group of European travelers who vanished in the Himalayas around that time."

When William Sax learned of the results, he was incredulous. He had spent years in the mountain villages below the lake, among the devotees of Nanda Devi. The women consider themselves to be keepers of the goddess's memory, and Sax had recorded and translated many of their songs and stories of the pilgrimage. He feels certain that if a large party of travelers, especially foreign travelers, had died at Roopkund in recent centuries, there would have been some record in folklore. After all, despite the new study's surprises, the Roopkund A group was not inconsistent with the earlier findings.

"I never heard a word, not a hint of a story, no folktale or anything," Sax told me. "And there's absolutely no reason to be up there if they *weren't* on the pilgrimage." The idea of a group of eighteenth-century Greeks on a Hindu pilgrimage seemed far-fetched. A simpler explanation would be that the Roopkund B bones somehow got mixed up while sitting in storage. "It is quite possible that these bones were contaminated," he said, and the researchers were simply taking their provenance on trust: "They didn't actually collect them themselves." Having been fascinated with the region's way of life for four decades, he also found the scientists' perspective lacking. "This isn't just a story about bones," he said. "It's also a story about human beings and religious devotion."

Many anthropologists and archaeologists are uneasy about the incursion of genomics into their domain and suspicious of its brash certainties. "We're not schooled in the nuances," Reich admitted to me. "Anthropologists and geneticists are two groups speaking different languages and getting to know each other." Research into human origins and the differences between populations is always vulnerable to misuse. The grim history of eugenics still casts a shadow over genetics—a field with limitless appeal for white supremacists and others looking to support racist views—even though, for half a century, geneticists have rejected the idea of large hereditary disparities among human populations for the great majority of traits. Genetic science was vital in discrediting racist biological theories and establishing that racial categories are ever-shifting social constructs that do not align with genetic variation. Still, some anthropologists, social scientists, and even geneticists are deeply uncomfortable with any research that explores the hereditary differences among populations. Reich is insistent that race is an artificial category rather than a biological one, but maintains that "substantial differences across populations" exist. He thinks that it's not unreasonable to investigate those differences scientifically, although he doesn't undertake such research himself. "Whether we like it or not, people *are* measuring average differences among groups," he said. "We need to be able to talk about these differences clearly, whatever they may be. Denying the possibility of substantial differences is not for us to do, given the scientific reality we live in."

In 2018, Reich published a book, *Who We Are and How We Got Here*, about how genetic science is revolutionizing our understanding of our species. After he presented material from the book as an op-ed in the *Times*, sixty-seven anthropologists, social scientists, and others signed an open letter on BuzzFeed, titled "How Not to Talk About Race and Genetics." The scholars complained that Reich's "skillfulness with ancient and contemporary DNA should not be confused with a mastery of the cultural, political, and biological meanings of human groups," and that Reich "critically misunderstands and misrepresents concerns" regarding the use of such loaded terms as "race" and "population."

Reich's lab now has an ethics-and-outreach officer, Jakob Sedig, whose job is to work with some of the cultural groups being studied, to understand and respond to their sensitivities. "We are mapping genetic groups to archaeological cultures," Sedig, who has a Ph.D. in anthropology from the University of Colorado, explained. "How we're defining these groups genetically is not how they see themselves culturally. We don't want to discredit other people's beliefs, but we don't want to censor our research based on those beliefs. There's no one answer. You need a dialogue from the beginning."

Reich acknowledges that geneticists need to be careful about how they discuss their work. He said that the majority of archaeologists and anthropologists welcome the insights that genetic research provides, although "there are a small number of Luddites who want to break our machines." In our conversations, Reich emphasized that the findings of geneticists were almost always unexpected and tended to explode stereotypes. "Again and again, I've found my own biases and expectations to be wrong," he said. "It should make us realize that the stories we tell ourselves about our past are often very different from the reality, and we should have humility about that." When I asked him for examples, he mentioned the origin of "white people"—light-skinned people from Europe and parts of western Asia. He assumed (as did most scientists) that whites represented a stable lineage that had spread across western Eurasia tens of thousands of years ago and established a relatively homogeneous population. But his research showed that as recently as eight thousand years ago there were at least four distinct groups of Europeans, as genetically different from one another as the British are from the Chinese today, some with brown skin color. As he put it in an email, "'White people' simply didn't exist ~8,000 years ago."

Around 500 B.C., the Greek traveler Scylax of Caryanda is said to have journeyed through parts of the Indian subcontinent and sailed down the Indus River. In his writings, known only from secondary sources, Scylax called the river Indos, from which the English name for the subcontinent derives. Alexander the Great invaded India in 326 B.C., having previously swept through what is now Afghanistan and Pakistan. His armies traversed the Indus plains and reached as far as the Beas River before turning back. There was lasting Hellenic influence in the region for centuries, although the eventual decline of Greek civilization largely brought direct contact with Greece to an end.

Perhaps, the Roopkund researchers thought, there might be a tribe or a group in India descended from Greeks. Alexander left behind commanders and soldiers in some of the territories he conquered, many of whom stayed. Members of the Kalash tribe, in northern Pakistan, claim to be descendants of Alexander's soldiers. (This was the inspiration for Rudyard Kipling's story "The Man Who Would Be King.") The Kalash are a distinct people with their own language and an ancient, animistic religion. Genetic research suggests that the Kalash have a western European origin, and one disputed study found Greek heritage. On investigation, Reich's team found that the modern genetic profile of the Kalash did not resemble that of Roopkund B. Two centuries before Christ, parts of northern India, Pakistan, and Afghanistan formed the Indo-Greek Kingdom, the easternmost state of the Hellenic world. But, again, Roopkund B didn't resemble any populations living there now.

Could Roopkund B have come from an *unsampled* population in India descended from Greeks or a related group? In this scenario, an enclave of migrants to India never admixed with South Asians, and retained their genetic heritage. But the genetics of Roopkund B, showing no sign of isolation or inbreeding, ruled this out, too. And then there was the stubborn fact that the Roopkund B people ate a diet more consistent with the Mediterranean than with India. The evidence pointed to one conclusion: they were Mediterranean travelers who somehow got to Roopkund, where they died in a single, terrible event. And yet the historians I consulted, specialists in South Asian and Greek history and authorities in the history of Himalayan mountaineering, said that, in recent centuries, there was no evidence of a large group of unrelated people from the eastern Mediterranean—men and women—traveling in the Himalayas before 1950.

Since the study was published, one of the most determined investigators of the mystery has been a recently retired archaeologist named Stuart Fiedel, whose main research focus is the migration of Paleo-Americans into the New World from Asia. "I hate unsolved mysteries," Fiedel told me. "It makes zero sense that a party of male and female Greek islanders would be participating in a Hindu pilgrimage around 1700 or 1800. That's because, one, there is no documented presence of any substantial Greek communities in northern India at those times, and, two, there is no record of Europeans converting to Hinduism or Buddhism in those periods."

He sent Harney and Reich a string of emails proposing alternatives to the Mediterranean theory. Fiedel contends that the mitochondrial DNA lineages and the Y-chromosomal DNA lineages of the Roopkund B group are rare or absent in the population of the Greek islands, but are relatively common in Armenians and other peoples of the Caucasus. His preferred hypothesis is that the Roopkund B people were Armenian traders. Armenians traveled widely in Tibet, India, and Nepal during the seventeenth and eighteenth centuries, trading in pearls, amber, and deer musk, a precious ingredient in perfume. Several large Indian cities have Armenian communities that go back centuries. "They might have been hanging with some major Hindu party trying to sell them stuff," Fiedel said. Noting that nothing of value was found on the bodies, he speculated that the travelers were killed by Thuggees, a cult of robbers and murderers whose fearsome reputation in British India gave us the word "thug." Thuggees were said to attach themselves to travelers or groups of pilgrims, gaining their trust and then robbing and murdering them on a remote stretch of road. "The Thuggees would make off with kids," Fiedel said. "Everybody in the Roopkund B population is mature. There isn't any gold on the skeletons, no rings, necklaces, or anklets on the victims. Who removed those things? And they were dumped in water. The Thuggees would dump people in water."

Reich and Harney reject Fiedel's genetic interpretation. Reich wrote back to him saying that the full DNA from Roopkund B was "extremely different from Armenians both modern and ancient." What's more, scholars increasingly view British reports about Thuggees as inaccurate or embellished, reflecting the colonialist fear and incomprehension of the country they occupied. Some historians question whether the Thuggees even existed.

Reich and Fiedel did agree, however, that Sax's suspicion that the bones could have simply been mixed up was unsustainable. A jumble of bones from a poorly curated storage area would not have the consistency of age, type, diet, and genetics displayed by the Roopkund B remains. The data would be all over the map. Besides, even if these bones were proved to have been mislabeled, that would merely create another mystery: How did a bunch of eighteenth-century Greek bones get into a storage vault in India?

For the time being, Roopkund holds its secrets, but it remains possible that an answer will eventually be found. Veena Mushrif-Tripathy, a bioarchaeologist on the previous study and a co-author of the new one, pointed out that Roopkund is so remote and inhospitable—in 2003, when she and her colleagues went to collect bones, altitude sickness forced her to turn back—there has never been a systematic archaeological investigation of the site. All the bones studied so far have been picked up haphazardly, a flawed way of sampling that often skews results. A careful excavation, she believes, might solve the mystery, especially if it is able to plumb the lake itself. The water is frozen most of the year, so the skeletons and artifacts visible on the lake bed have been kept safe from looters and souvenir hunters. "Inside the lake, you can get more preserved bones with soft tissues," she said. "And if they are Greek people we should get some artifacts or tools or something which we can trace back to Greece."

And what of Nanda Devi? The new study established that multiple groups had died at the lake centuries apart. Did everyone die in hailstorms? Mushrif-Tripathy thinks that a hailstorm was probably involved in one mass death but that most people had likely just died of exposure. According to Ayushi Nayak, who performed the isotopic bone analyses at the Max Planck Institute, Hindu pilgrims sometimes go barefoot and thinly clothed to sacred sites in the Himalayas as a spiritual challenge. Completing the pilgrimage in this way is a sign that the goddess favors you and wants you to survive. In other words, most of the Roopkund dead probably perished as Sax almost did, when he was an undergraduate—staggering around in a sudden blizzard and looking for their companions.

UPDATE

When this article was first assigned to me in 2019, I made plans to hike to Roopkund Lake—a difficult, high-altitude trek that takes about a week. Unfortunately, the pandemic intervened and prevented my trip. As a journalist, I've always believed it is essential to visit key localities in the story, to carry the reader along and make the writing as vivid and true as possible. I also feel this is important when writing a novel—to bring the reader into the setting. With the Roopkund mystery I wasn't able to do this, and as a result the article had to be recast into a more standard journalistic format.

It was a shame—if I had gone to Roopkund Lake, experienced the Himalayas, and seen the skeletons in person, the article would have been quite different—more dramatic and interesting.

I expected that after the New Yorker *piece was published, someone with specialized knowledge of Greek pilgrims in India might come forward with an answer to the mystery. But no one did, and no progress has been made in unraveling the enigma of the (probable) Greek skeletons at Roopkund Lake—who they might have been and why they were there.*

THE SKIERS AT DEAD MOUNTAIN

IGOR DYATLOV WAS a tinkerer, an inventor, and a devotee of the wilderness. Born in 1936, near Sverdlovsk (now Yekaterinburg), he built radios as a kid and loved camping. When the Soviet Union launched Sputnik, in 1957, he constructed a telescope so that he and his friends could watch the satellite travel across the night sky. By then, he was an engineering student at the city's Ural Polytechnic Institute. One of the leading technical universities in the country, U.P.I. turned out top-flight engineers to work in the nuclear-power and weapons industries, communications, and military engineering. During his years there, Dyatlov led a number of arduous wilderness trips, often using outdoor equipment that he had invented or improved. It was a time of optimism in the U.S.S.R. Khrushchev's Thaw had freed many political prisoners from Stalin's Gulag, economic growth was robust, and the standard of living was rising. The shock that the success of Sputnik delivered to the West further bolstered national confidence. In late 1958, Dyatlov began planning a winter expedition that would exemplify the boldness and vigor of a new Soviet generation: an ambitious sixteen-day cross-country ski trip in the Urals, the north-south mountain range that divides western Russia from Siberia, and thus Europe from Asia.

Originally published in the *New Yorker* in 2021 as "Has an Old Soviet Mystery at Last Been Solved?"

He submitted his proposal to the U.P.I. sports club, which readily approved it. Dyatlov's itinerary lay three hundred and fifty miles north of Sverdlovsk, in the traditional territory of the Mansi, an indigenous people. The Mansi came into contact with Russians around the sixteenth century, when Russia was extending its control over Siberia. Though largely Russified by this time, the Mansi continued to pursue a semi-traditional way of life—hunting, fishing, and reindeer herding. Dyatlov's group would ski two hundred miles, on a route that no Russian, as far as anyone knew, had taken before. The mountains were gentle and rounded, their barren slopes rising from a vast boreal forest of birch and fir. The challenge wouldn't be rugged terrain but brutally cold temperatures, deep snow, and high winds.

Dyatlov recruited his classmate Zina Kolmogorova and seven other fellow students and recent graduates. They were among the elite of Soviet youth and all highly experienced winter campers and cross-country skiers. One was Dyatlov's close friend Georgy Krivonishchenko, who had graduated from U.P.I. two years before and worked as an engineer at the Mayak nuclear complex, in the then secret town of Chelyabinsk-40. Jug-eared, small, and wiry, he told jokes, sang, and played the mandolin. Two other recent graduates were Rustem Slobodin and Nikolay Thibault-Brignoles, of French descent, whose father had been worked nearly to death in one of Stalin's camps. The other students included Yuri Yudin, Yuri Doroshenko, and Aleksandr Kolevatov. The youngest of the group, at twenty, was Lyuda Dubinina, an economics major, a track athlete, and an ardent Communist, who wore her long blond hair in braids tied with silk ribbons. On a previous wilderness outing, Dubinina had been accidentally shot by a hunter, and survived—quite cheerfully, it was said—a fifty-mile journey back to civilization. A couple of days before the group was due to set off, the U.P.I. administration unexpectedly added a new member, much older than the others and largely unknown to them: Semyon Zolotaryov, a thirty-seven-year-old veteran of the Second World War with an old-fashioned mustache, stainless-steel crowns on his teeth, and tattoos.

The party left Sverdlovsk by train on January 23. Several of them hid under seats to avoid buying tickets. They were in high spirits—so high that on a layover between trains Krivonishchenko was briefly detained by police for playing his mandolin and pretending to panhandle in the train station. We know these details because there was a communal journal, and many of the skiers also kept personal journals. At least five had cameras, and the pictures they took show a lively and strikingly handsome group of young people having the adventure of their lives—skiing, laughing, playing in the snow, and mugging for the camera.

After two days on trains, the party reached Ivdel, a remote town with a Stalin-era prison camp that, by then, held mostly criminals. From there the group traveled another day by bus, then in the back of a woodcutter's truck, and finally by ski, guided by a horse-drawn sleigh. They slept in an abandoned logging camp called Second Northern. There Yuri Yudin had a flareup of sciatica that forced him to pull out of the trip. The next day, January 28, he turned back, while the remaining nine set off toward the mountains. The plan was to end up at the tiny village of Vizhai around February 12, and telegram the U.P.I. sports club that they had arrived safely. The expected telegram never came.

AT FIRST, THE U.P.I. sports club assumed that the group had just been held up; there had been reports of a heavy snowstorm in the mountains. But after several days passed, families of the group began placing frantic phone calls to the university and to the local bureau of the Communist Party, and, on February 20, a search was launched. There were several search parties: student volunteers from U.P.I., prison guards from the Ivdel camp, Mansi hunters, local police; the military deployed planes and helicopters. On February 25, the students found ski tracks, and the next day they discovered the skiers' tent—above the tree line on a remote mountain that Soviet officials referred to as Height 1079 and that the Mansi called Kholat Syakhl, or Dead Mountain. There was no one inside.

The tent was partly collapsed and largely buried in snow. After digging it out, the search party saw that the tent appeared to have been deliberately slashed in several places. Yet, inside, everything was neat and orderly. The skiers' boots, axes, and other equipment were arranged on either side of the door. Food was laid out as if about to be eaten; there was a stack of wood for a heating stove, and clothes, cameras, and journals.

About a hundred feet downhill, the search party found "very distinct" footprints of eight or nine people, walking (not running) toward the tree line. Almost all the prints were of stockinged feet, some even bare. One person appeared to be wearing a single ski boot. "Some of the prints indicated that the person was either barefoot or in socks because you could see the toes," a searcher later testified. The party followed the prints downhill for six to seven hundred yards, until they vanished near the tree line.

The next morning, searchers found the bodies of the mandolin player Krivonishchenko and the student Doroshenko under a tall cedar tree at the edge of the forest. They were lying next to a dead fire, wearing only underwear. Twelve to fifteen feet up the tree were some recently broken branches, and on the trunk bits of skin and torn clothes were found. Later that day, a search party discovered the bodies of Dyatlov and Kolmogorova. Both were farther up the slope, facing in the direction of the tent, their fists tightly clenched. They seemed to have been trying to get back there.

The four bodies were autopsied, while the search for the others continued. The medical examiner noted a number of bizarre features. Krivonishchenko had blackened fingers and third-degree burns on a shin and a foot. Inside his mouth was a chunk of flesh that he had bitten off his right hand. Doroshenko's body had burned hair on one side of the head and a charred sock. All the bodies were covered with bruises, abrasions, scratches, and cuts, as was a fifth body, that of the recent graduate Slobodin, which was discovered a few days later. Like Dyatlov and Kolmogorova, Slobodin was on the slope leading back to the tent, with a sock on one foot and a felt bootie on the other; his autopsy noted a minor fracture to his skull.

By now, a homicide investigation was under way, led by a prosecutor in his mid-thirties named Lev Ivanov. Toxicology tests were done, witness testimony taken, diagrams and maps made of the scene, and evidence gathered and forensically analyzed. The tent and its contents were helicoptered out of the mountains and set up again inside a police station. This led to a key discovery: a seamstress who came to the station to do a uniform fitting happened to notice that the slashes in the tent had been made from the inside.

Something had happened that induced the skiers to cut their way out of the tent and flee into the night, into a howling blizzard, in twenty-below-zero temperatures, in bare feet or socks. They were not novices to the winter mountains; they would have been acutely aware of the fatal consequences of leaving the tent half dressed in those conditions. This is the central, and apparently inexplicable, mystery of the incident.

Four bodies remained missing. In early May, when the snow began to melt, a Mansi hunter and his dog came across the remains of a makeshift snow den in the woods two hundred and fifty feet from the cedar tree: a floor of branches laid in a deep hole in the snow. Pieces of tattered clothing were found strewn about: black cotton sweatpants with the right leg cut off, the left half of a woman's sweater. Another search team arrived and, using avalanche probes around the den, they brought up a piece of flesh. Excavation uncovered the four remaining victims, lying together in a rocky streambed under at least ten feet of snow. The autopsies revealed catastrophic injuries to three of them. Thibault-Brignoles's skull was fractured so severely that pieces of bone had been driven into the brain. Zolotaryov and Dubinina had crushed chests with multiple broken ribs, and the autopsy report noted a massive hemorrhage in the right ventricle of Dubinina's heart. The medical examiner said the damage was similar to what is typically seen as the "result of an impact of an automobile moving at high speed." Yet none of the bodies had external penetrating wounds, though Zolotaryov's was missing its eyes, and Dubinina's was missing its eyes, tongue, and part of the upper lip.

A careful inventory of clothing recovered from the bodies revealed that some of these victims were wearing clothes taken or cut off the bodies of others, and a laboratory found that several items emitted unnaturally high levels of radiation. A radiological expert testified that, because the bodies had been exposed to running water for months, these levels of radiation must originally have been "many times greater."

On May 28, Ivanov abruptly closed the investigation. His role was to determine whether a crime had been committed, not to clarify what had happened, and he concluded that homicide was not a factor. Ivanov ended his report with a non-explanation that has bedeviled Dyatlov researchers ever since: "It should be concluded that the cause of the hikers' demise was an overwhelming force, which they were not able to overcome."

In classic Soviet style, a number of officials who had little to do with the tragedy were either punished or fired, including the director of U.P.I. and the chairman of its sports club, the local Communist Party secretary, the chairmen of two workers' unions, and a union inspector. The investigative files, photographs, and journals were classified, and the area around Dead Mountain was placed off-limits to skiers and outdoor enthusiasts for years. The tent was stored but eventually became moldy and had to be thrown out. The saddle in the mountains which the skiers were heading for but never reached was named the Dyatlov Pass.

The victims' families were left deeply dissatisfied. Many of them wrote to officials, including Khrushchev, demanding a more thorough investigation. But nothing more was done, and the mysterious deaths of the nine skiers subsided into relative obscurity.

IN 1990, THE prosecutor Ivanov, who had retired, published an article in which he claimed that, while compiling his 1959 report, he'd been pressured not to include his views on what happened. The article, titled "The Enigma of the Fireballs," said that the skiers had been killed by heat rays or balls of fire associated with U.F.O.s. In his original examination of the scene, Ivanov had found trees with unusual burn marks, which "confirmed that some kind of heat ray, say, or a powerful force whose nature is completely unknown (to us, at least) acted selectively on specific objects"—in this case, people. The last photograph in Krivonishchenko's camera showed flares and streaks of light against a black background.

By then, the official files had been released, and in the decades since, the case has become one of the most celebrated mysteries of the Soviet era. It has generated dozens of books and documentaries, along with a slew of websites and message boards on which Dyatlov obsessives trade scores of theories—the official count of the Russian Prosecutor General's office lists seventy-five—about what happened. In 2000, relatives and friends of the victims established the Dyatlov Group Memorial Foundation, whose purpose is to honor the memory of the skiers and seek the truth. Its president is Yuri Kuntsevich, who, as a twelve-year-old boy, attended the funerals of some of the victims. He went on to study and teach at U.P.I. (which has since become the Ural State Technical University) and to join its sports club. Now in his mid-seventies, he still leads tours to the Dyatlov Pass. Kuntsevich told me that Russians generally favor one of two theories: the skiers died because they had stumbled into an area where secret weapons were being tested; alternatively, the party was "killed by mercenaries," probably American spies.

Kuntsevich insists that the first of these theories is the correct one, and it's also what the families tend to believe. The idea is that a missile launch of some kind went disastrously wrong, inflicting severe injuries on some of the skiers and forcing the group to flee their tent, at which point they either froze to death or were killed by military observers. Yuri Yudin, whose sciatica compelled him to abandon the trip, likewise maintained that the deaths were not natural. Not long before he died, in 2013, he declared that his teammates had been taken from the tent at gunpoint and murdered. Dubinina, he said, may have had her tongue cut out by the killers because she was the most outspoken of the group.

Proponents of the weapons-test theory cite claims from people in the region that they had seen flashes of light or moving balls of fire in the direction of the mountains. In 2008, a three-foot-long piece of metal was found in the area; according to the Dyatlov Foundation, which took possession of it, the metal is part of a Soviet ballistic missile. Military tests would explain the radioactivity of recovered clothing. Yevgeny Okishev, Ivanov's supervisor in the Prosecutor General's office, gave an interview to a newspaper in 2013, in which he recalled finding it suspicious when he and his colleagues were instructed to test recovered items for radiation. He sent a letter to his superiors asking why radiation was relevant. In response, the Deputy Prosecutor General met with the team. Okishev said that the official dodged questions about weapons testing and ordered them to tell people that the deaths were accidental. "The victims' parents came to my office, some screamed and called us Fascists for hiding the truth from them," Okishev recalled. "But the case was closed, and not on our orders."

The theory, however, is not consistent with what was found at the site. There was no evidence that other people had been there. Snow does not lie: it would have been close to impossible to erase signs of the people and equipment involved in killing the group and restaging the scene. Besides, why make the staging so elaborate and bizarre? Why scatter the bodies around the landscape, cut off the clothing of some and dress others in it, build a snow den, bury four bodies in ten feet of snow, light a fire, and climb a tree to break branches, leaving skin on the bark? The theory would also suggest that there was a secret weapons base in the area, or that an errant missile had exploded over it. Yet despite the mass declassification of documents from the Soviet era and the diligent searches of Dyatlov enthusiasts, no such evidence has emerged.

The K.G.B. theory centers on Zolotaryov, the man who was foisted on the group at the last minute. A book published in Russia claims that he and two other skiers were K.G.B. agents on an assignment to meet with a group of C.I.A. operatives, to furnish them with deliberately misleading information. Samples of clothing contaminated by radioactive isotopes were to be offered as bait; the C.I.A. agents discovered the deception, killed them, and staged the scene. It is certainly possible that Zolotaryov had a K.G.B. link. His service record in the Second

World War had holes and inconsistencies, and his sudden inclusion certainly seems suspicious. Still, a K.G.B. connection, even if proved, wouldn't mean much; many people were low-level informants at the time. And the idea that the C.I.A. would have chosen a place like Dead Mountain for a rendezvous strains credulity.

Another class of theories considers a variety of natural disasters. An avalanche, perhaps, struck the tent, causing the crushing injuries to three of the victims and forcing the whole group to cut their way out and head to the forest for shelter. But no avalanche debris was found—a ski pole holding up the front of the tent was still standing—and the original investigation determined that the slope was too shallow to generate an avalanche. Besides, the injuries to the three victims found in the streambed were totally incapacitating. They could never have made it there unassisted—it was more than a mile from the tent—but the tracks leading downhill showed no signs of anyone being dragged. There were eight or nine separate sets of footprints, so the fatal injuries must have come after everyone had left the tent. A 2013 bestseller by the filmmaker and writer Donnie Eichar suggests that high winds passing over the mountain created infrasound, vibrations below the range of human hearing, and that this induced such terror that the skiers fled. Much about the book is excellent—Eichar conducted many interviews in Russia and traveled to the Dyatlov Pass in winter—but his thesis would require all nine people to have been so terrified of a sound they couldn't even hear that they ran to certain death, not grabbing their coats or boots, and slashing their way out when the tent door would have made for a far easier exit.

Various hypotheses considered in the 1959 inquest have also been raked over: carbon monoxide poisoning from the heater; sudden madness caused by consuming bad alcohol or hallucinogenic mushrooms that the Mansi sometimes hung on trees to dry; or even murder by the Mansi themselves, if, for instance, the party had strayed onto sacred land. But the autopsies ruled out the first two of these, and when the original investigators interviewed the local Mansi they found them "well disposed toward Russians," and believable. The Mansi had provided valuable help in the search, and they told the investigators that the area was not sacred; on the contrary, it was considered windy, barren, and worthless.

By far the most entertaining theory is that the party was attacked by a yeti. The final photograph found in Thibault-Brignoles's camera has become famous: a dark figure advancing through the snowy forest, hunched and menacing, with no facial features. The Discovery Channel built an entire show, *Russian Yeti: The Killer Lives*, around the image. The skiers actually had been joking about yetis a few hours before they died. A spoof propaganda leaflet was found in the tent. Alongside such items as "Greeting the XXI Congress with increased birthrate among hikers" was the following: "Science: In recent years there has been a heated debate about the existence of the Yeti. Latest evidence indicates that the Yeti lives in the northern Urals, near Mount Otorten." Still, the photograph, though blurry, pretty clearly shows a member of the group. Similarly, the Krivonishchenko image of streaks of light, which has been used to bolster the U.F.O. and weapons-test theories, is typical of the end of a film roll.

ALL THE DYATLOV theories share a basic assumption that the full story has not been told. In a place where information has been as tightly controlled as in the former Soviet Union, mistrust of official narratives is natural, and nothing in the record can explain why people would leave a tent undressed, in near-suicidal fashion. For decades, the families and the Dyatlov Group Memorial Foundation pressed for a new investigation; in 2019, elderly relatives of several victims finally succeeded in getting the case reopened.

A young prosecutor in Yekaterinburg, Andrei Kuryakov, was put in charge. In 2019, he organized a winter expedition to the site. His team took measurements, surveyed, photographed, and conducted a variety of experiments. Using photogrammetry of the pictures taken in 1959, they tried to establish the precise location of the tent. The spot they settled on was several hundred feet from a cairn marking the previously accepted location, on a steeper section of Kholat Syakhl's slope. Combing through historical data, the investigators determined that weather conditions on the mountain that night were even more extreme than had been thought. The skiers were engulfed in a storm with winds of up to sixty-five miles an hour and temperatures around minus thirty degrees Fahrenheit. As evening fell, they were probably unsure of their precise location.

From the outset, Kuryakov adopted an intentionally narrow scope, dismissing seventy-two of the seventy-five explanations for what may have happened. "A large class of these seventy-five versions are conspiracy theories alleging that the authorities were somehow involved in the incident," he said, when announcing the investigation. "We have already proved that this is absolutely false." This left the investigation with three natural occurrences to consider: an avalanche, a hurricane, and a slab of snow sliding over the tent. Last July [of 2020], Kuryakov held a televised press conference in which he told his audience that the last of these was the definitive explanation.

Two photographs taken by the Dyatlov party at around five p.m., while they pitched the tent, show that they cut deeply into the snowpack at right angles to the slope, forming a hollow. They had picked a spot where the mountain peak offered some shelter from the strongest winds. Later in the evening, Kuryakov said, a snow slab detached from the slope above and buried most of the tent, pinning down the occupants and possibly causing injuries. Fearing that a full-scale avalanche was imminent, the skiers cut their way out of the downslope side of the tent and fled to a rock ridge a hundred and fifty feet away, which Kuryakov termed a "natural avalanche limiter." But the big avalanche didn't come, and, in pitch darkness, they were unable to find their way back to the tent and took shelter in the woods, a mile away. Kuryakov tested this theory by blindfolding a man and a woman and leading them ninety feet downhill from a tent. Asked to find their way back, they quickly went astray. The task would have been even more difficult in a blizzard, with most of the tent buried in snow.

Analyzing 1959 photographs, many Dyatlov researchers had calculated that the tent was pitched on a slope of some fifteen degrees, which is not steep enough to sustain the movement of snow in cold conditions. The new position of the tent as determined by Kuryakov's topographical experts was therefore crucial, because the gradient here was between twenty-three and twenty-six degrees, enough for avalanche formation. A paper corroborating much of Kuryakov's explanation was published in January 2021 by two Swiss engineers in the journal *Communications Earth & Environment*. Creating a mathematical model of the snow structure that night, the researchers showed why the slab didn't release immediately when the group cut into it, but only hours later: additional loading of snow during the storm was responsible.

I reviewed the hypothesis with Ethan Greene, the director of the Colorado Avalanche Information Center, who has a Ph.D. in the physics of heat and mass transfer in snow. He suggested that the party's decision to pitch the tent in the wind shadow of the peak made it likely that they were cutting into a so-called wind slab—an accumulation of hard snow even more dangerous than a typical snow slab. Compacted by the wind, this kind of snow is several times denser than directly deposited snow and, according to Greene, can weigh as much as six hundred and seventy pounds per cubic yard. Furthermore, the clear conditions preceding the storm could have led to the formation of a layer of light, feathery frost, known as surface hoar. When buried in fresh snow during the storm, this layer forms a hazardous stratum that provides poor support to the snow above and often releases, resulting in avalanches. By removing the support on the lower edge of the slab while digging to set their tent, the skiers likely caused it to fracture higher up.

If the wind slab had simply slid over the tent and halted, without developing into a full-fledged avalanche, the evidence, Greene said, might not be visible twenty-five days later. Even the fissure in the snowpack would probably have been erased by the elements. If a three- to four-foot-thick slab moved over the tent, each skier's body would have been covered by as much as a thousand pounds. The massive weight prevented them from retrieving their boots or warm clothing and forced them to cut their way out of the downslope side of the tent.

The two Swiss researchers believe that the snow slab probably caused the terrible injuries to three of the skiers found at the snow den, but this remains unlikely, given the distance of those bodies from the tent. Kuryakov's explanation was more ingenious. The nine skiers retreated downhill, taking shelter under the cedar tree and building a fire. Because the young trees nearby were icy and wet, someone climbed the cedar to break branches higher up—hence the skin and scraps of clothing found on the trunk. The fire they built, in these extreme conditions, was not enough to save them, however. The two most poorly dressed of the group died first. The burned skin on their bodies came from their desperate efforts to seek warmth from the fire. This would suggest that the piece of flesh Krivonishchenko bit from his finger was probably a result of the delirium that overtakes someone who's dying of hypothermia, or perhaps from an attempt to test for sensation in a frostbitten hand.

The surviving skiers cut the clothes off their dead comrades and dressed themselves in the remnants. At some point, the group split up. Three skiers, including Dyatlov, tried to return to the tent and soon froze to death as they struggled uphill. The other four, who were better dressed, decided to build a snow den to shelter in overnight. They needed deep snow, which they found in a ravine a couple of hundred feet away. Unfortunately, the spot they picked lay above a stream, a tributary of the Lozva River. The stream, which never freezes, had hollowed out a deep icy tunnel, and the group's digging caused its roof to collapse, throwing them onto the rocky streambed and burying them in ten to fifteen feet of snow. The pressure of tons of snow forcing them against the rocks caused the traumatic injuries found in this group. The gruesome facial damage—the missing tongue, eyes, and lip—probably resulted from scavenging by small animals and from decomposition.

Kuryakov's reconstruction of events made a single plausible narrative out of previously mystifying anomalies. But what of the radiation? This detail, the most enigmatic of all, might be the easiest to explain. For one thing, the mantles used in camp lanterns at the time contained small amounts of the radioactive element thorium. Even more pertinent, the expedition took place less than two years after the world's third-worst nuclear accident (after Chernobyl and Fukushima), which occurred at the Mayak nuclear complex, south of Sverdlovsk, in September 1957. A tank of radioactive waste exploded and a radioactive plume some two hundred miles long—later named the East Urals Radioactive Trace—spread northward. Krivonishchenko had worked at the facility and helped with the cleanup, and another skier came from a village in the contaminated zone.

KURYAKOV CLOSED HIS press conference by declaring, "Formally, this is it. The case is closed." Given how freighted the case is in Russia, this was too optimistic. For many people, nature alone cannot explain a tragedy of this magnitude; perpetrators must be identified and the state and its dark past invoked. Sure enough, the conclusions were greeted with scorn, especially by the families of the dead. The Dyatlov Group Memorial Foundation sent a letter to the Prosecutor General declaring that, in its view, the skiers' deaths were caused by "the atmospheric release of a powerful toxic substance" when a secret weapons test went wrong. Natalia Varsegova, a Moscow journalist, who has covered the subject for many years, also rejected Kuryakov's conclusions. "Two years ago I thought that the prosecutor Andrei Kuryakov really wanted to know the truth," she wrote to me in an email. "But now I doubt it. I don't believe in an avalanche." After the Swiss report came out, she published an article rejecting it as well. "These theoreticians' conclusions are supported by mathematical calculations, formulas, and diagrams, but the local Mansi, numerous tourists, and organizers of snowmobile tours, who have never seen avalanches on this slope, are unlikely to agree with them."

A month after the press conference, Kuryakov was reprimanded for holding it without authorization, and in October he was removed from his post. (The prosecutor's office has claimed that he resigned, and he did not respond to requests for an interview.) He was then appointed a deputy minister of natural resources in the Sverdlovsk region, which is a major timber producer. As Kuntsevich wrote to me sarcastically, Kuryakov was shunted off to "felling trees." Meanwhile, the Prosecutor General declined to be interviewed for this article, and his office has issued no official report. Kuntsevich believes that a report may never be released, even to the families. The foundation is now calling for yet another investigation. Any clarity that Kuryakov's solution might have brought was quickly occluded amid an atmosphere of murk and mistrust.

The most appealing aspect of Kuryakov's scenario is that the Dyatlov party's actions no longer seem irrational. The snow slab, according to Greene, would probably have made loud cracks and rumbles as it fell across the tent, making an avalanche seem imminent. Kuryakov noted that although the skiers made an error in the placement of their tent, everything they did subsequently was textbook: they conducted an emergency evacuation to ground that would be safe from an avalanche, they took shelter in the woods, they started a fire, they dug a snow cave. Had they been less experienced, they might have remained near the tent, dug it out, and survived. But avalanches are by far the biggest risk in the mountains in winter, and the more experience you have, the more you fear them. The skiers' expertise doomed them.

At the end of 1958, as the date of departure approached, Krivonishchenko wrote a letter to Dyatlov firming up various logistical matters, and he enclosed a poem addressing New Year's greetings to the entire group:

> *Here's wishing you*
> *Camps pitched on mounts afar,*
> *Routes to hike over ranges untamed,*
> *Packs that, as ever, rest lightly on your backs,*
> *And weather that smiles upon your quest. . . .*
> *And let your footprints*
> *Trace winding tracks across the map of Russia.*

Today, the Dyatlov Pass is a popular hiking and tourist destination. Hundreds have visited Height 1079, and followed Dyatlov's route on foot, snowmobile, or skis. People come from all over the world to see the place where the tent once stood, the streambed where bodies were found, and the cedar tree, its broken branches still visible. Others come to take measurements, photographs, and videos to support their pet theories. The windswept heights of Dead Mountain have become a site of pilgrimage. Long after their deaths, Dyatlov and his friends did indeed leave their footprints across the map of Russia.

UPDATE

After the article was published, a well-known production studio offered to purchase an option on the article to make a television series from it. After months of negotiations I received the contract in the mail, but my name had been written incorrectly, so I sent it back for that minor correction. They sent me a corrected copy, which arrived on the day before Russia invaded Ukraine—and the production company immediately withdrew the offer. This turned out to be fortuitous, because my writing partner Lincoln Child and I decided to transform the story into a thriller ourselves, featuring our two characters Special Agent Corinne Swanson and archaeologist Nora Kelly. Dead Mountain *was published in August 2023.*

Since the publication of the article, no further information of note has come to light about the Dyatlov mystery, although far-fetched theories continue to proliferate. I firmly believe the disgraced Russian prosecutor solved the mystery once and for all, but few in Russia—or anywhere else for that matter—accept his conclusions.

THE SKELETON ON THE RIVERBANK

ON SUNDAY, JULY 28, 1996, in the middle of the afternoon, two college students who were watching a hydroplane race on the Columbia River in Kennewick, Washington, decided to take a shortcut along the river's edge. While wading through the shallows, one of them stubbed his toe on a human skull partly buried in the sand. The students picked it up and, thinking it might be that of a murder victim, hid it in some bushes and called the police.

Floyd Johnson, the Benton County coroner, was called in, and the police gave him the skull in a plastic bucket. Late in the afternoon, Johnson called James Chatters, a forensic anthropologist and the owner of a local consulting firm called Applied Paleoscience. "Hey, buddy, I got a skull for you to look at," Johnson said. Chatters had often helped the police identify skeletons and distinguish between those of murder victims and those found in Indian burial sites. He is a small, determined, physically powerful man of forty-eight who used to be a gymnast and a wrestler. His work occasionally involves him in grisly or spectacular murders, where the victims are difficult to identity, such as burnings and dismemberments.

"When I looked down at the skull," Chatters told me, "right off the bat I saw it had a very large number of Caucasoid features"—in particular, a long, narrow braincase, a narrow face, and a slightly projecting upper jaw. But when Chatters took it out of the bucket and laid

---

Originally published in the *New Yorker* in 1997 as "The Lost Man."

it on his worktable he began to see some unusual traits. The crowns of the teeth were worn flat, a common characteristic of prehistoric Indian skulls, and the color of the bone indicated that it was fairly old. The skull sutures had fused, indicating that the individual was past middle age. And, for a prehistoric Indian of what was then an advanced age, he or she was in exceptional health; the skull had, for example, all its teeth and no cavities.

As dusk fell, Chatters and Johnson went out to the site to see if they could find the rest of the skeleton. There, working in the dying light, they found more bones, lying around on sand and mud in about two feet of water. The remains were remarkably complete: only the sternum, a few rib fragments, and some tiny hand, wrist, and foot bones were missing. The bones had evidently fallen out of a bank during recent flooding of the Columbia River.

The following day, Chatters and Johnson spread the bones out in Chatters's laboratory. In forensic anthropology, the first order of business is to determine sex, age, and race. Determining race was particularly important, because if the skeleton turned out to be Native American it fell under a federal law called the Native American Graves Protection and Repatriation Act, or NAGPRA. Passed in 1990, NAGPRA requires the government—in this case, the Army Corps of Engineers, which controls the stretch of the Columbia River where the bones were found—to ascertain if human remains found on federal lands are Native American and, if they are, to "repatriate" them to the appropriate Indian tribe.

Chatters determined that the skeleton was male, Caucasoid, from an individual between forty and fifty-five years old, and about five feet nine inches tall—much taller than most prehistoric Native Americans in the Northwest. In physical anthropology, the term "Caucasoid" does not necessarily mean "white" or "European"; it is a descriptive term applied to certain biological features of a diverse category that includes, for example, some South Asian groups as well as Europeans. (In contrast, the term "Caucasian" is a culturally defined racial category.) "I thought maybe we had an early pioneer or fur trapper," he said. As he was cleaning the pelvis, he noticed a gray object embedded in the bone, which had partly healed and fused around it. He took the bone to be X-rayed, but the object did not show up, meaning that it was not made of metal. So he requested a CAT scan. To his surprise, the scan revealed the object to be part of a willow-leaf-shaped spear point, which had been thrust into the bone and broken off. It strongly resembled a Cascade projectile point—an Archaic Indian style in wide use from around nine thousand to forty-five hundred years ago.

The Army Corps of Engineers asked Chatters to get a second opinion. He put the skeleton into his car and drove it a hundred miles to Ellensburg, Washington, where an anthropologist named Catherine J. MacMillan ran a forensic consulting business called the Bone-Apart Agency. "He didn't say anything," MacMillan told me. "I examined the bones, and I said, 'Male, Caucasian.' He said, 'Are you sure?' and I said, 'Yeah.' And then he handed me the pelvis and showed me the ancient point embedded in it, and he said, 'What do you think now?' And I said, 'That's extremely interesting, but it still looks Caucasian to me.'" In her report to the Benton County coroner's office she wrote that in her opinion the skeleton was "Caucasian male."

Toward the end of the week, Chatters told Floyd Johnson and the Army Corps of Engineers that he thought they needed to get a radiocarbon date on Kennewick Man. The two parties agreed, so Chatters sent the left fifth metacarpal bone—a tiny bone in the hand—to the University of California at Riverside.

On Friday, August 23, Jim Chatters received a telephone call from the radiocarbon lab. The bone was between ninety-three hundred and ninety-six hundred years old. He was astounded. "It was just a phone call," he said. "I thought, Maybe there's been a mistake. I had to see the report with my own eyes." The report came on Monday, in the form of a fax. "I got very nervous then," Chatters said. He knew that, because of their age, the bones on his worktable had to be one of the most important archaeological finds of the decade. "It was just a tremendous responsibility." The following Tuesday, the coroner's office issued a press release on the find, and it was reported in the *Seattle Times* and other local papers.

Chatters called in a third physical anthropologist, Grover S. Krantz, a professor at Washington State University. Krantz looked at the bones on Friday, August 30. His report noted some characteristics common to both Europeans and Plains Indians but concluded that "this skeleton cannot be racially or culturally associated with any existing American Indian group." He also wrote, "The Native Repatriation Act has no more applicability to this skeleton than it would if an early Chinese expedition had left one of its members there."

Fifteen minutes after Krantz finished looking at the bones, Chatters received a call from Johnson. Apologetically, the coroner said, "I'm going to have to come over and get the bones." The Army Corps of Engineers had demanded that all study of the bones cease, and had required him to put the skeleton in the county sheriff's evidence locker. On the basis of the carbon date, the Corps had evidently decided that the skeleton was Native American and that it fell under NAGPRA.

"When I heard this, I panicked," Chatters said. "I was the only one who'd recorded any information on it. There were all these things I should have done. I didn't even have photographs of the postcranial skeleton. I thought, Am I going to be the last scientist to see these bones?"

On September 9, the Umatilla Indians, leading a coalition of five tribes and bands of the Columbia River basin, formally claimed the skeleton under NAGPRA, and the Corps quickly made a preliminary decision to "repatriate" it. The Umatilla Indian Reservation lies just over the border, in northeastern Oregon, and the other tribes live in Washington and Idaho; all consider the Kennewick area part of their traditional territories. The Umatillas announced that they were going to bury the skeleton in a secret site, where it would never again be available to science.

Three weeks later, the *New York Times* picked up the story, and from there it went to *Time* and on around the world. Television crews from as far away as France and Korea descended on Kennewick. The Corps received more than a dozen other claims for the skeleton, including one from a group known as the Asatru Folk Assembly, the

California-based followers of an Old Norse religion, who wanted the bones for their own religious purposes.

On September 2, the Corps had directed that the bones be placed in a secure vault at the Pacific Northwest National Laboratory, in Richland, Washington. Nobody outside of the Corps has seen them since. They are now at the center of a legal controversy that will likely determine the course of American archaeology.

WHAT WAS A Caucasoid man doing in the New World more than ninety-three centuries ago? In the reams of press reports about the discovery, that question never seemed to be dealt with. I called up Douglas Owsley, who is the Division Head for Physical Anthropology at the National Museum of Natural History, Smithsonian Institution, in Washington, D.C., and an expert on Paleo-American remains. I asked how many well-preserved skeletons that old had been found in North America.

He replied, "Including Kennewick, about seven."

Then I asked if any others had Caucasoid features, and there was a silence that gave me the sense that I was venturing onto controversial ground.

He guardedly replied, "Yes."

"How many?"

"Well," he said, "in varying degrees, all of them."

Kennewick Man's bones are part of a growing quantity of evidence that the earliest inhabitants of the New World may have been a Caucasoid people. Other, tentative evidence suggests that these people may have originally come from Europe. The new evidence is fragmentary, contradictory, and controversial. Critical research remains to be done, and many studies are still unpublished. At the least, the new evidence calls into question the standard Beringian Walk theory, which holds that the first human beings to reach the New World were Asians of Mongoloid stock, who crossed from Siberia to Alaska over a land bridge. The new evidence involves three basic questions. Who were the original Americans? Where did they come from? And what happened to them?

"You're dealing with such a black hole," Owsley told me. "It's hard to draw any firm conclusions from such a small sample of skeletons, and there is more than one group represented. That's why Kennewick is so important."

KENNEWICK MAN MADE his appearance at the dawn of a new age in physical anthropology. Scientists are now able to extract traces of organic material from a person's bone and perform a succession of biochemical assays which can reveal an astonishing amount of information about the person. In the late 1990s, for example, scientists at Oxford University announced that they had compared DNA extracted from the molar cavity of a nine-thousand-year-old skeleton known as Cheddar Man to DNA collected from fifteen pupils and five adults from old families in the village of Cheddar, in Somersetshire. They had established a blood tie between Cheddar Man and a schoolteacher who lived just half a mile from the cave where the bones were found.

In the few weeks that Kennewick Man was in the hands of scientists, they discovered a great deal about him. Isotopic-carbon studies of the bones indicate that he had a diet high in marine food—that he may have been a fisherman who ate a lot of salmon. He seems to have been a tall, good-looking man, slender and well proportioned. (Studies have shown that "handsomeness" is largely the result of symmetrical features and good health, both of which the Kennewick Man had.) Archaeological finds of similar age in the area suggest that he was part of a small band of people who moved about, hunting, fishing, and gathering wild plants. He may have lived in a simple sewn tent or a mat hut that could be disassembled and carried. Some nearby sites contain large numbers of fine bone needles, indicating that a lot of delicate sewing was going on: Kennewick Man may have worn tailored clothing. For a person at that time to live so long in relatively good health indicates that he was clever or lucky, or both, or had family and close friends around him.

He appears to have perished from recurring infections caused by the stone point in his hip. Because of the way his bones were found, and the layer of soil from which they presumably emerged, it may be that he was not deliberately buried but died somewhere near the river and was swept away and covered up in a flood. He may have perished alone on a fishing trip, far from his family.

Chatters made a cast of the skull before the skeleton was taken from his office. In the months since, he has been examining it to figure out how Kennewick Man may have looked. He plans to work with physical anthropologists and a forensic sculptor to make a facial reconstruction. "On the physical characteristics alone, he could fit on the streets of Stockholm without causing any kind of notice," Chatters told me. "Or on the streets of Jerusalem or New Delhi, for that matter. I've been looking around for someone who matches this Kennewick gentleman, looking for weeks and weeks at people on the street, thinking, This one's got a little bit here, that one a little bit there. And then, one evening, I turned on the TV and there was Patrick Stewart"—Captain Picard, of *Star Trek*—"and I said, 'My God, there he is! Kennewick Man!'"

IN SEPTEMBER, FOLLOWING the requirements of NAGPRA, the Corps advertised in a local paper its intention of repatriating the skeleton secreted in the laboratory vault. The law mandated a thirty-day waiting period after the advertisements before the Corps could give a skeleton to a tribe.

Physical anthropologists and archaeologists around the country were horrified by the seizure of the skeleton. They protested that it was not possible to demonstrate a relationship between nine-thousand-year-old remains and any modern tribe of the area. "Those tribes are relatively new," says Dennis Stanford, the chairman of the Department of Anthropology at the Smithsonian's National Museum of Natural History. "They pushed out other tribes that were there." Both Owsley and Richard L. Jantz, a biological anthropologist at the University of Tennessee, wrote letters to the Army Corps of Engineers in late September saying that the loss to science would be incalculable if Kennewick Man were to be reburied before being studied. They received no response. Robson Bonnichsen, the director of the Center for the Study of the First Americans, at Oregon State University, also wrote to the Corps and received no reply. Three representatives and a United States senator from the state of Washington got in touch with the Corps, pleading that it allow the skeleton to be studied before reburial, or, at least, refrain from repatriating the skeleton until Congress could take up the issue. The Corps rebuffed them.

The Umatillas themselves issued a statement, which was written by Armand Minthorn, a tribal religious leader. Minthorn, a small, well-spoken young man with long braids, is a member of a new generation of Native American activists, who see religious fundamentalism—in this case, the Washat religion—as a road back to Native American traditions and values:

> Our elders have taught us that once a body goes into the ground, it is meant to stay there until the end of time.... If this individual is truly over 9,000 years old, that only substantiates our belief that he is Native American. From our oral histories, we know that our people have been part of this land since the beginning of time. We do not believe that our people migrated here from another continent, as the scientists do.... Scientists believe that because the individual's head measurement does not match ours, he is not Native American. Our elders have told us that Indian people did not always look the way we look today. Some scientists say that if this individual is not studied further, we, as Indians, will be destroying evidence of our history. We already know our history. It is passed on to us through our elders and through our religious practices.

Despite the mounting protests, the Corps refused to reconsider its decision to ban scientific study of the Kennewick skeleton. As the thirty-day waiting period came to a close, anthropologists around the country panicked. Just a week before it ended, on October 23, a group of eight anthropologists filed suit against the Corps. The plaintiffs included Douglas Owsley, Robson Bonnichsen, and also Dennis Stanford. Stanford, one of the country's top Paleo-Indian experts, is a formidable opponent. While attending graduate school in New Mexico, he roped in local rodeos, and helped support his family by leasing an alfalfa farm. There's still a kind of laconic, frontier toughness about him. "Kennewick Man has the potential to change the way we view the entire peopling of the Americas," he said to me. "We had to act. Otherwise, I might as well retire."

The eight are pursuing the suit as individuals. Their academic institutions are reluctant to get involved in a lawsuit as controversial as this, particularly at a time when most of them are negotiating with tribes over their own collections.

In the suit, the scientists have argued that Kennewick Man may not meet the NAGPRA definition of "Native American" as being "of, or relating to a tribe, people, or culture that is indigenous to the United States." The judge trying the case has asked both sides to be prepared to define the word "indigenous" as it is used in NAGPRA. This will be an interesting exercise, since no human beings are indigenous to the New World: we are all immigrants.

The scientists have also argued that the Corps had had no evidence to support its claim that the skeleton had a connection to the Umatillas. Alan Schneider, the scientists' attorney, says, "Our analysis of NAGPRA is that first you have to make a determination if the human remains are Native American. And then you get to the question of cultural affiliation. The Army Corps assumed that anyone who died prior to a certain date is automatically Native American."

The NAGPRA law appears to support the scientists' point of view. It says that when there are no known lineal descendants, "cultural affiliation" should be determined using "geographical, kinship, biological, archaeological, anthropological, linguistic, folkloric, oral traditional, historical," or other "relevant information or expert opinion" before human remains are repatriated. In other words, human remains must often be studied before anyone can say whom they are related to.

The Corps, represented by the Justice Department, has refused to comment on most aspects of the case. "It's really as if the government didn't want to know the truth about Kennewick Man," Alan Schneider told me in late April. "It seems clear that the government will *never* allow this skeleton to be studied, for any reason, unless it is forced to by the courts."

Preliminary oral arguments in the case were heard on June 2 in the United States District Court for the District of Oregon. The scientists asked for immediate access to study the bones, and the Corps asked for summary judgment. Judge John Jelderks denied both motions, and said that he would have a list of questions for the Corps that it is to answer within a reasonable time. With the likelihood of appeals, the case could last a couple of years longer, and could ultimately go to the Supreme Court.

Schneider was not surprised that the Corps had sided with the Indians. "It constantly has a variety of issues it has to negotiate with Native American tribes," he told me, and he specified, among others, land issues, water rights, dams, salmon fishing, hydroelectric projects, and toxic-waste dumps. The Corps apparently decided, Schneider speculated, that in this case its political interests would be better served by supporting the tribes than by supporting a disgruntled group of anthropologists with no institutional backing, no money, and no political power. There are large constituencies for the Indians' point of view: fundamentalist Christians and liberal supporters of Indian rights. Fundamentalists of all varieties tend to object to scientific research into the origins of humankind, because the results usually contradict their various creation myths. A novel coalition of conservative Christians and liberal activists was important in getting NAGPRA through Congress.

KENNEWICK MAN, EARLY as he is, was not one of the first Americans. But he could be their descendant. There is evidence that those mysterious first Americans were a Caucasoid people. They may have come from Europe and may be connected to the Clovis people of

America. Kennewick may provide evidence of a connection between the Old World and the New.

The Clovis mammoth hunters were the earliest widespread culture that we know of in the Americas. They appeared abruptly, seemingly out of nowhere, all over North and South America about eleven thousand five hundred years ago—two thousand years before Kennewick. (They were called Clovis after a town in New Mexico near an early site—a campground beside an ancient spring which is littered with projectile points, tools, and the remains of fires.) We have only a few fragments of bone from the Clovis people and their immediate descendants, the Folsom, and these remains are so damaged that nothing can be learned from them at present.

The oldest bones that scientists have been able to study are the less than a dozen remains that are contemporaneous with Kennewick. They date from between eight thousand and nearly eleven thousand years ago—the transition period between the Paleo-Indian and the Archaic Indian traditions. Most of the skeletons have been uncovered accidentally in recent years, primarily because of the building boom in the West. (Bones do not survive well in the East: the soil is too wet and acidic.) Some other ancient skeletons, though, have been discovered gathering dust in museum drawers. Among these oldest remains are the Spirit Cave mummy and Wizard's Beach Willie, both from Nevada; the Hourglass Cave and Gordon's Creek skeletons, from Colorado; the Buhl Burial, from Idaho; and remains from Texas, California, and Minnesota.

Douglas Owsley and Richard Jantz made a special study of several of these ancient remains. The best-preserved specimen they looked at was a partial mummy from Spirit Cave, Nevada, which is more than nine thousand years old. Owsley and Jantz compared the Spirit Cave skull with thirty-four population samples from around the world, including ten Native American groups. In an as yet unpublished letter to the Nevada State Museum, they concluded that the Spirit Cave skull was "very different" from any historic-period Native American groups. They wrote, "In terms of its closest classification, it does have a 'European' or 'Archaic Caucasoid' look, because morphometrically it is most similar to the Ainu from Japan and a medieval period Norse population." Additional early skeletons that they and others have looked at also show Caucasoid-like traits that, in varying degrees, resemble Kennewick Man's. Among these early skeletons, there are no close resemblances to modern Native Americans.

But even though the skeletons do look Caucasoid, other evidence indicates that the concept of "race" may not be applicable to human beings of ten or fifteen thousand years ago. Recent studies have discovered that all Eurasians may have looked Caucasoid-like in varying degrees. In addition, some researchers believe that the Caucasoid type first emerged in western Asia or the Middle East, rather than in Europe. The racial differences we see today may be a late (and trivial) development in human evolution. If this is the case, Kennewick Man may indeed be a direct ancestor of today's Native Americans—an idea that some preliminary DNA and dental studies seem to support.

BIOLOGY, HOWEVER, ISN'T the whole story. There is some archaeological evidence that the Clovis people of America—Kennewick Man's predecessors—came from Europe, which could account for his Caucasoid features. When the Clovis people appeared in the New World, they possessed an advanced stone and bone technology, and employed it in hunting big game. (It is no small feat to kill a mammoth or a mastodon with a handheld spear or an atlatl.) If the Clovis people—or their precursors—had migrated to North America from Asia, one would expect to find early forms of their distinctive tools, of the right age, in Alaska or eastern Siberia. We don't. But we do find such artifacts in Europe and parts of Russia.

Bruce Bradley is the country's leading expert on Paleo-Indian flaked-stone technology. In 1970, as a recent college graduate, Bradley spent time in Europe studying Paleolithic artifacts, including those of the Solutrean people, who lived in southwestern France and Spain between twenty thousand and sixteen thousand years ago. During his stay in Europe, Bradley learned how to flake out a decent Solutrean point. Then he came back to America and started studying stone tools made by the Clovis people. He noticed not only striking visual similarities between the Old World Solutrean and the New World Clovis artifacts, as many researchers had before, but also a strikingly similar use of flaking technology.

"The artifacts don't just look identical," Bradley told me. "They are *made* the same way." Both the European Solutrean and the American Clovis stone tools are fashioned using the same complex flaking techniques. He went on, "As far as I know, overshot flaking is a technique unique to Solutrean and Clovis, and the only diving flaking that is more than eleven thousand years old is that of Solutrean and Clovis."

To argue his case, he drove from Cortez, Colorado, where he lives, to see me in Santa Fe, bringing with him a trunkful of Solutrean artifacts, casts of Clovis bone and stone tools, and detailed illustrations of artifacts, along with a two-pound chunk of gray Texas flint and some flaking tools.

Silently, he laid a piece of felt on my desk and put on it a cast of a Clovis knife from Blackwater Draw, New Mexico. On top of that, he put a broken Solutrean knife from Laugerie-Haute, in France. They matched perfectly—in size, shape, thickness, and pattern of flaking. As we went through the collection, the similarities could be seen again and again. "It isn't just the flaking as you see it on the finished piece that is the same," Bradley explained. "Both cultures had a very specific way of preparing the edge before striking it to get a very specific type of flake. I call these 'deep technologies.' These are not mere resemblances—they are deep, complex, abstract concepts applied to the stone."

I remarked that according to other archaeologists I had talked to the resemblances were coincidental, the result of two cultures—one Old World, the other New World—confronting the same problems and solving them in the same way. "Maybe," Bradley said. "But a lot of older archaeologists were not trained with technology in mind. They see convergence of form or look, but they don't see the technology that goes into it. I'm not saying this is the final answer. But there is so much similarity that we cannot say, 'This is just a coincidence,' and ignore it."

To show me what he meant, he brought out the piece of gray flint, and we went into the backyard. "This isn't just knocking away at a stone," he said, hefting the piece of flint and examining it from various angles with narrowed eyes. "After the initial flaking of a spear point, there comes a stage where you have almost limitless choices on how to continue." He squatted down and began to work on the flint with hammer stones and antler billets, his hands deftly turning and shaping the material—knapping, chipping, scraping, pressing, flaking. The pile of razor-sharp flakes grew bigger, and the chunk of flint began to take on a definite shape. Over the next ninety minutes, Bradley reduced the nodule to a five-inch-long Clovis spear point.

It was a revelation to see that making a Clovis point was primarily an intellectual process. Sometimes ten minutes would pass while Bradley examined the stone and mapped out the next flaking sequence. "After thirty years of intense practice, I'm still at the level of a mediocre Clovis craftsman," he said, wiping his bloodied hands and reviving himself with a cup of cocoa. "This is as difficult and complex as a game of chess."

Those are not the only similarities between the Old World Solutrean and the New World Clovis cultures. As we went through Bradley's casts of Clovis artifacts, he compared them with pictures of similar Solutrean artifacts. The Clovis people, for example, produced enigmatic bone rods that were beveled on both ends and crosshatched; so did the Solutrean. The Clovis people fashioned distinctive spear points out of mammoth ivory; so did the Solutrean. Clovis and Solutrean shaft wrenches (tools thought to have been used for straightening spears) look almost identical. At the same time, there are significant differences between the Clovis and the Solutrean tool kits; the Clovis people fluted their spear points, for instance, while the Solutrean did not.

"To get to the kind of complexity you find in Clovis tools, I see a long technological development," Bradley said. "It isn't one person in one place inventing something. But there is no evolution in Clovis technology. It just appears, full blown, all over the New World, around eleven thousand five hundred years ago. Where's the evolution? *Where did that advanced Clovis technology come from?*"

When I mentioned the idea of a possible Old World Solutrean origin for the New World Clovis to Lawrence Straus, a Solutrean expert at the University of New Mexico, he said, "There are two gigantic problems with it—thousands of years of separation and thousands of miles of ocean." Pointing out that the Solutrean technology itself appeared relatively abruptly in the South of France, he added, "I think this is a fairly clear case of mankind's ability to invent things."

But Bradley and other archaeologists have pointed out that these objections may not be quite so insurmountable. Recently, in southern Virginia, archaeologists discovered a layer of non-Clovis artifacts beneath a Clovis site. The layer dates from fifteen thousand years ago—much closer to the late Solutrean period, sixteen thousand five hundred years ago. Only a few rough tools have been found, but if more emerge they might provide evidence for the independent development of Clovis—or provide a link to the Solutrean. The as yet unnamed culture may be precisely the precursor to Clovis that archaeologists have been looking for since the 1930s. The gap between Solutrean and Clovis may be narrowing.

The other problem is how the Solutrean people—if they are indeed the ancestors of the Clovis—might have reached America in the first place. Although most of them lived along riverbanks and on the seacoast of France and Spain, there is no evidence that they had boats. No Paleo-Indian boats have been found on the American side either, but there is circumstantial evidence that the Paleo-Indians used boats. The ancestors of the Australian aborigines got to Australia in boats from the Indonesian archipelago at least fifty thousand years ago.

But the Solutrean people may not have needed boats at all: sixteen thousand years ago, the North Atlantic was frozen from Norway to Newfoundland. Seasonal pack ice probably extends as far south as Britain and Nova Scotia. William Fitzhugh, the director of the Arctic Studies Center, at the Smithsonian, points out that, if human beings had started in France, crossed the English Channel, then hopped along the archipelago from Scotland to the Faroes to Iceland to Greenland to Newfoundland, and, finally, Nova Scotia, the biggest distance between landfalls would have been about five hundred miles, which Fitzhugh says could have been done by foot over ice. It would still have been a stupendous journey, but perhaps not much more difficult than the Beringian Walk, across thousands of miles of tundra, muskeg, snow, and ice.

"For a long time, most archaeologists have been afraid to challenge the Beringian Walk paradigm," Bradley said. "I don't want to try and convince anybody. But I do want to shake the bushes. You could put all the archaeological evidence for the Asian-Clovis connection in an envelope and mail it for thirty-two cents. The evidence for a European-Clovis connection you'd have to send in a UPS box, at least."

Robson Bonnichsen, the director of the Center for the Study of the First Americans, at Oregon State University, is another archaeologist whose research is challenging the established theories. "There is a presumption, written into almost every textbook on prehistory, that Paleo-Americans such as Clovis are the direct ancestors of today's Native Americans," he told me. "But now we have a very limited number of skeletons from that early time, and it's not clear that that's true. We're getting some hints from people working with genetic data that these earliest populations might have some shared genetic characteristics with latter-day European populations. A lot more research is needed to sort all this out. Now, for the first time, we have the technology to do this research, especially in molecular biology. Which is why we *must* study Kennewick."

This summer [1997], Bonnichsen hopes to go to France and recover human hair from Solutrean and other Upper Paleolithic sites. He will compare DNA from that hair with DNA taken from naturally shed Paleo-American hair recovered from the United States to see if there is a genetic link. Human hair can survive thousands of years in the ground, and, using new techniques, Bonnichsen and his research team have been finding hair in the places where people worked and camped.

BONNICHSEN AND MOST other archaeologists tend to favor the view that if the ancestors of Clovis once lived in Europe they came to America via Asia—the Beringian Walk theory with somewhat different people doing the walking. C. Vance Haynes Jr., the country's top Paleo-Indian geochronologist, who is a professor of anthropology at the University of Arizona, and is a plaintiff in the lawsuit, said to me, "When I look at Clovis and ask myself where in the world the culture was derived from, I would say Europe." In an article on the origins of Clovis, Haynes noted that there were extraordinary resemblances between New World Clovis and groups that lived in Czechoslovakia and Ukraine twenty thousand years ago. He noted at least nine "common traits" shared by the Clovis and certain eastern European cultures: large blades, end scrapers, burins, shaft wrenches, cylindrical bone points, knapped bone, unifacial flake tools, red ochre, and circumferentially chopped mammoth tusks. He also pointed out that an eighteen-thousand-year-old burial site of two children near Lake Baikal, in Central Asia, exhibits remarkable similarities to what appears to be a Clovis burial site of two cremated children in Montana. The similarities extend beyond tools and points buried with the remains: red ochre, a kind of iron oxide, was placed in both graves. This suggests a migratory group carrying its technology from Europe across Asia. "If you want to speculate, I see a band moving eastward from Europe through Siberia, and meeting people there, and having cultural differences," Haynes said to me. "Any time there's conflict, it drives people, and maybe it just drove them right across the Bering land bridge. And exploration could have been as powerful a driving force thirteen thousand years ago as it was in 1492."

ONCE THE CLOVIS people or their predecessors reached the New World, what happened to them? This is the second—and equally controversial—half of the theory: that the Clovis people or their immediate successors, the Folsom people, may have been supplanted by the ancestors of today's Native Americans. In this scenario, Kennewick Man may have been part of a remnant Caucasoid population related to Clovis and Folsom. Dennis Stanford, of the Smithsonian, said to me, "For a long time, I've held the theory that the Clovis and the Folsom were overwhelmed by a migration of Asians over the Bering land bridge. It may not just have been a genetic swamping or a pushing aside. The north Asians may have been carrying diseases that the Folsom and the Clovis had no resistance to"—just as European diseases wiped out a large percentage of the Native American population after the arrival of Columbus. Stanford explained that at several sites the Paleo-Indian tradition of Clovis and Folsom was abruptly replaced by Archaic Indian traditions, which had advanced but very different lithic technologies. The abruptness of the transition and the sharp change in technology, Stanford feels, suggest a rapid replacement of Folsom by the Archaic cultural complex, rather than an evolution from one into the other. The Archaic spear point embedded in Kennewick Man's hip could even be evidence of an ancient conflict: the Archaic Indian tradition was just beginning to appear in the Pacific Northwest at the time of Kennewick Man's death.

OWSLEY AND OTHER physical anthropologists who have studied the skulls of the earliest Americans say that the living population they most closely match is the mysterious Ainu, the aboriginal inhabitants of the Japanese islands. Called the Hairy People by the Japanese, the Ainu are considered by some researchers to be a Caucasoid group who, before mixing with the Japanese in the late nineteenth and twentieth centuries, had European faces, wavy hair, thick beards, and a European-type distribution of body hair. Early travelers reported that some also had blue eyes. Linguists have not been able to connect the Ainu language with any other on earth. The American Museum of Natural History, in New York, has a collection of nineteenth-century photographs of pure-blooded Ainu, which I have examined: they stare from the glass plates like fierce, black-bearded Norwegians.

Historically, the Ainu have been the "Indians" of Japan. After the ancestors of the Japanese migrated from the mainland a couple of thousand years ago, they fought the Ainu and pushed them into the northernmost islands of the Japanese archipelago. The Japanese later discriminated against the Ainu, forcing their children to attend Japanese schools and suppressing their religion and their language. Today, most Ainu have lost their language and many of their distinct physical characteristics, although there has recently been a movement among them to recapture their traditions, their religion, their language, and their songs. Like the American Indians, the Ainu suffer high rates of alcoholism. The final irony is that the Ainu, like the American Indians for Americans, have become a popular Japanese tourist attraction. Many Ainu now make a living doing traditional dances and selling handicrafts to Japanese tourists. The Japanese are as fascinated by the Ainu as we are by the Indians. The stories are mirror images of each other, with only the races changed.

If the Ainu are a remnant population of those people who crossed into America thirteen or more millennia ago, then they are right where one might expect to find them, in the extreme eastern part of Asia. Stanford says, "That racial type goes all the way to Europe, and I suspect that originally they were the same racial group at both ends. At that point in time, this racial diversification hadn't developed. They could have come into the New World from two directions at once, east and west."

THERE IS A suspicion among anthropologists that some of the people behind the effort to rebury Kennewick and other ancient skeletons are afraid that the bones could show that the earliest Americans were Caucasoid. I asked Armand Minthorn about this. "We're not afraid of the truth," he said calmly. "We already know our truth. We're not telling the scientists what *their* truth is."

The Umatillas were infuriated by the research that Chatters did on the skeleton before it was seized by the Corps. "Scientists have dug up and studied Native Americans for decades," Minthorn wrote. "We view this practice as desecration of the body and a violation of our most deeply held religious beliefs." Chatters told me that he had received "vitriolic" and "abusive" telephone calls from tribe members, accusing him of illegalities and racism. (The latter was an odd charge, since Chatters's wife is of Native American descent.) A client of his, he said, received an unsigned letter from one of the tribes, telling her not to work with Chatters anymore. "They're going to ruin my livelihood," the forensic consultant said.

In a larger sense, the anger of the Umatillas and other Native American tribes is understandable, and even justified. If you look into the acquisition records of most large, old natural-history museums, you will see a history of unethical, and even grisly, collecting practices. Fresh graves were dug up and looted, sometimes in the dead of night. "It is most unpleasant to steal bones from a grave," the eminent anthropologist Franz Boas wrote in his diary just around the turn of the century, "but what is the use, someone has to do it." Indian skulls were bought and sold among collectors like arrowheads and pots. Skeletons were exhibited with no regard for tribal sensitivities. During the Indian Wars, warriors who had been killed on the battlefield were sometimes decapitated by Army doctors so that scientists in the East could study their heads. The American Museum of Natural History "collected" six live Inuits in Greenland and brought them back to New York to study; four of them died of respiratory diseases, whereupon the museum macerated their corpses and installed the bones in its collections. When I worked in the museum, in the 1980s, entire hallways were lined with glass cases containing Indian bones and mummified body parts—a small fraction of the museum's collection, which includes an estimated twenty thousand or more human remains, of all races. Before NAGPRA, the Smithsonian had some thirty-five thousand sets of human remains in storage; around eighteen thousand of them were Native American.

Now angry Native Americans, armed with NAGPRA and various state reburial laws, are emptying such museums of bones and grave goods. Although most anthropologists agree that burials identified with particular tribes should be returned, many have been horrified to discover that some tribes are trying to get everything—even skeletons and priceless funerary objects that are thousands of years old.

An amendment that was introduced in Congress last January [1997] would tighten NAGPRA further. The amended law could have the effect of hindering much archaeology in the United States involving human remains, and add to the cost of construction projects that inadvertently uncover human bones. (Or perhaps the law would merely guarantee that such remains would be quietly destroyed.)

Native Americans have already claimed and reburied two of the earliest skeletons, the Buhl Burial and the Hourglass Cave skeleton, both of which apparently had some Caucasoid characteristics. The loss of the Buhl Burial was particularly significant to anthropologists, because it was more than ten thousand years old—a thousand years older than Kennewick Man—and had been found buried with its grave goods. The skeleton, a woman between eighteen and twenty years old, received, in the opinion of some anthropologists, inadequate study before it was turned over to the Shoshone-Bannock tribe. The Northern Paiute have asked that the Spirit Cave mummy be reburied. If these early skeletons are all put back in the ground, anthropologists say, much of the history of the peopling of the Americas will be lost.

WHEN DARWIN PROPOSED his theory of natural selection, it was seized upon and distorted by economists, social engineers, and politicians, particularly in England: they used it to justify all sorts of vicious social and economic policies. The scientific argument about the original peopling of the Americas threatens to be distorted in a similar way. Some tabloids and radio talk shows have referred to Kennewick as a "white man" and have suggested that his discovery changes everything with respect to the rights of Native Americans in this country. James Chatters said to me, "There are some less racially enlightened folks in the neighborhood who are saying, 'Hey, our ancestors were here first, so we don't owe the Indians anything.'"

This is clearly racist nonsense: these new theories cannot erase or negate the existing history of genocide, broken treaties, and repression. But it does raise an interesting question: If the original inhabitants of the New World were Europeans who were pushed out by Indians, would it change the Indians' position in the great moral landscape?

"No," Stanford said in reply to this question. "Whose ancestors are the people who were pushed out? And who did the pushing? The answer is that we're all the descendants of those folks. If you go back far enough, eventually we all have a common ancestor—*we're all the same*. When the story is finally written, the peopling of the Americas will turn out to be far more complicated than anyone imagined. There have been a lot of people who came here, at many different times. Some stayed and some left, some made it and some didn't, some got pushed out and some did the pushing. It's the history of humankind: the tough guy gets the ground."

Chatters put it another way. "We didn't go digging for this man. He fell out—he was actually a volunteer. I think it would be wrong to stick him back in the ground without waiting to hear the story he has to tell. We need to look at things as human beings, not as one race or another. The message this man brings to us is one of unification: there may be some commonality in our past that will bring us together."

UPDATE

In the legal struggle over whether scientists would be allowed to study Kennewick Man, the scientists prevailed. Douglas Owsley, a physical anthropologist at the Smithsonian Institution, and a team of scientists intensively studied the remains and hypothesized, based on the morphology of the cranium, that Kennewick Man was only distantly related to present-day Native Americans of the area. The cranial features, they concluded, most closely resembled Polynesian or East Asian peoples. Their research also showed, based on an isotopic analysis of the teeth and bones, that Kennewick Man was a wanderer whose place of origin was much farther north than the Pacific Northwest region. They also concluded he had been deliberately interred. At the same time, efforts were made to extract and analyze Kennewick Man's DNA, a very difficult task. That effort was finally successful in 2015, when the University of Copenhagen in Denmark determined that, contrary to the previous multiple scientific analyses, Kennewick Man was "very closely related" to the local tribes of the area. In 2016, Congress passed legislation to return the remains to the local tribes, and the bones were reburied in an undisclosed location on February 18, 2017. The question of who the first Americans were and where they came from, however, remains open and is the subject of fierce debate.

UNSOLVED MYSTERIES

THE MYSTERY OF OAK ISLAND

ON OAK ISLAND, everybody gets up early. By dawn, with the fog turning into a drizzle, the crew is hard at work. I've taken refuge inside the rusted hulk of an old tank car, where I can take notes without the ink smearing. Up the hill, men cluster around a drilling rig that is pounding its way into the island's interior. Shouts and curses echo through the fog. "Sand!" someone yells. "We're in sand, damn it!"

All around me lies the evidence of the hunt: big Ingersoll-Rand air compressors, enormous pump heads, piles of steel casing, acetylene tanks, strange infernal machines, and bright aluminum ducts snaking their way across the ground. The chill September rain is slowly coating them all.

In 1909, a young law clerk named Franklin Delano Roosevelt trod this very ground with pick, shovel, and high hopes. Admiral Richard Byrd, Errol Flynn, and Vincent Astor all at one time or another took an interest.

Here in Mahone Bay, about forty miles southwest of Halifax, Nova Scotia, I am at the site of the most intensive treasure hunt in history, a hunt that has lasted 193 years, cost millions of dollars, and killed six men. The raison d'être for it all is a narrow, water-filled shaft called the Money Pit—and what may be hidden in its muddy depths. To date, not one penny of treasure has been recovered. Nor does anyone

---

Originally published in *Smithsonian* magazine in June 1988 as "Death Trap Defies Treasure Seekers for Two Centuries."

know what might be buried here, who buried it, or why. The island stubbornly refuses to yield anything but the most tantalizing and infuriatingly ambiguous clues.

But the Oak Island mystery may soon be solved. Triton Alliance Ltd., a group of Canadian and American investors, is making the biggest assault yet on the Money Pit. They are digging a shaft of gargantuan proportions twenty stories into the very heart of the island. In doing so, Triton will either find treasure and uncover an important archaeological site, or they will have burned up $10 million digging an empty hole.

The mystery of Oak Island began in the summer of 1795, when a teenage farm boy named Daniel McGinnis decided to do a little exploring. He rowed out to Oak Island, tied up his boat, and started poking around. His story, along with those of the many who have followed, goes something like this:

At the seaward end of the island, the thick forest of red oaks suddenly gave way to an old clearing, dotted with a few rotted stumps. In the center stood an ancient oak with a sawed-off limb. The limb showed evidence of rope burns and, in some versions of the tale, had an old ship's tackle hanging from it. Directly underneath, the ground had subsided into a shallow depression. From this, a young boy could draw only one conclusion: buried pirate treasure.

McGinnis returned the next day with two friends, Anthony Vaughan and John Smith, and they began digging. At two feet they struck a tier of flagstones. On pulling these up, they found themselves digging in what appeared to be an old shaft excavated in the hard glacial till, a mixture of clay, sand, gravel, and rocks. The shaft had been filled with loose dirt and they could see old pick marks in the walls.

At about ten feet they hit a platform of rotten logs, the ends embedded in the clay. They eagerly ripped these up and kept going. At twenty feet they struck another platform, and yet another at thirty. With no end in sight, and no doubt their chores seriously in arrears, the three boys gave up—but only for the time being. Both McGinnis and Smith later bought land on the island, hoping eventually to reach the vast treasure that they were sure must lie at the bottom of the pit.

When a well-to-do man named Simeon Lynds heard the story, he enlisted workers, including the three young men, in a new assault. Work began in 1803. At forty feet they struck another log platform. They continued to hit platforms at regular intervals, and they also encountered a layer of charcoal, a layer of putty, and a layer of fibrous material that was later identified as coconut fiber.

At ninety feet they found something really exciting—a flat stone inscribed with mysterious figures. They quickly tore up the platform beneath it. Soon, water began seeping into the pit and they found themselves bailing as much as digging. As night came on, they probed the muck at the bottom of the pit with a crowbar and struck something hard at the ninety-eight-foot level.

"Some supposed it was wood," one researcher wrote later, "and others called it a chest. This circumstance put them all in good spirits and during the evening a good deal of discussion arose as to who should have the largest share of the treasure."

There would be no sharing of treasure. The next day the diggers arose to find the pit sixty feet deep in water—*salt* water. Bailing proved to be as futile as bailing out the ocean.

This first, failed effort was only the beginning. Syndicate after syndicate was floated to get to the bottom of the pit. They dug, pumped, excavated, drilled, dynamited, trenched, cribbed, bulldozed, and blasted the island, turning the eastern end into a cratered wasteland. At some point in the early nineteenth century the original hole was nicknamed the "Money Pit," although the only direction money seemed to go was into the pit, not out of it.

In 1849 diggers built a platform over the Money Pit and cored down with a pod auger, a primitive type of drill. The drilling engineer, Jotham B. McCully, later stated that the drill struck wood at ninety-eight feet, dropped through twelve inches of space, then rattled through "twenty-two inches of metal in pieces," struck more wood, another twenty-two inches of metal, then wood, then soil. The auger failed to bring up any metal except three links of a gold chain which, McCully theorized, "had apparently been forced from an epaulette."

Around this time, treasure hunters made another curious discovery. One day a workman was sitting alongside the cobbled beach at Smith's Cove, a small cove five hundred feet east of the Money Pit. He noticed that as the tide ebbed, the beach "gulched forth water like a sponge being squeezed." The crew immediately built a cofferdam around the spot and excavated the beach. To their astonishment, they discovered that the beach was a fake—that is, it had been made to look like a beach but was in fact a giant filtering and drainage system. Underneath the cobbles they found thick layers of eel grass and coconut fiber lying on top of an elaborate system of box drains. The drains led, like the five fingers of a hand, to a point opposite the Money Pit. They were, apparently, the head of a "flood trap" designed to keep the pit filled with water.

The reader may well wonder how the original diggers intended to retrieve their treasures from such a death trap. Current theories, for which there is yet no evidence, are convincingly simple. Once the pirates—let us call them that for the moment—had dug the Money Pit sufficiently deep, they would have started side shafts that sloped gently back toward the surface. Treasures would have been hidden in the ends of these side tunnels, three hundred to five hundred feet away from the Money Pit but perhaps only thirty feet below the surface. The pirates would have known the direction and distance from the Money Pit, left highly visible as a decoy, to each of the treasure troves and it would have been a simple matter to dig them up.

In 1897, drillers brought up more strange clues. From the 155-foot level, the drill bit carried up a half-inch-square piece of parchment with two letters written on it with a quill pen. In another hole the drill was stopped cold at 126 feet by what seemed to be an iron plate. A magnet was raked through grit brought up from the hole and it pulled out thousands of iron filings. A year later, dye dumped into the pit emerged from the seabed at Smith's Cove, providing more evidence of a tunnel connection. But it also emerged from the South Shore Cove, establishing the existence of two flood tunnels, thus making things more complicated.

EVERYONE ASSUMED THAT whoever would go to that much trouble must have buried an enormous treasure. Around the turn of the century, fortune hunters estimated it at $10 million; by the 1930s, this had doubled; by the sixties, some people were talking about $100 million or more. Today it is pegged at $500 million to "several billion."

So what has been the problem? Why in the world hasn't someone been able to get to the bottom of the Money Pit?

The blame can be laid squarely on the treasure hunters themselves. Until Triton took over, each digger had believed he was almost there and worked in a frenzy. Many syndicates kept no records. Important artifacts were thrown away, lost, or destroyed. Drillers so churned the ground that much of the really significant evidence was obliterated to a depth of 150 feet.

That's not all. In 1861, so much digging had been going on that the bottom dropped out—literally. One weekend the diggers heard a crash echoing out of the Money Pit and rushed over in time to see the bottom of the pit drop into a void. Then, as they watched in horror, 10,000 board feet of cribbing unraveled and sank into the opening; shortly thereafter the pit itself caved in with another loud thump. Everything had vanished into an underground morass.

As if that weren't enough, the water seemed unstoppable. You could dig a shaft as deep as you wished, but as soon as you angled it toward the Money Pit, bang! the water burst through and it was every man for himself.

Worst of all, the treasure hunters managed to lose the Money Pit itself. So many pits, tunnels, and shafts had been dug that eventually nobody remembered exactly where the original was.

Of all the stories surrounding Oak Island, the one about the severed hand is by far the strangest. It was seen in a water-filled cavity at the bottom of a shaft known as Borehole 10X. The cavity was found during test drilling in the late 1960s; a narrow shaft was sunk to explore it further. In 1971, Dan Blankenship had enlarged Borehole 10X to the point where he could finally fit an underwater video camera down through it. He was monitoring the screen in a nearby shack while three crew members manned the equipment outside. The camera shortly came to rest in the cavern. There was a moment of silence. And then the crew heard a bloodcurdling yell from the shack.

"I called in each man," Dan recalls, "one at a time. I didn't say anything, just pointed to the screen. And each man said, 'Damn, that's a hand. That's a human hand.' The hand appeared to be floating in perfect equilibrium in the water."

Come on. A human hand?

Dan looks me straight in the eye. "Now I don't say I *think* I saw a human hand in there. I don't say that. I *saw* a hand. There's no question about it."

Once again, Oak Island had thrown up a maddening, intriguing clue. After a while you start asking yourself, What is real? How do you separate fact from fiction? Where's the truth?

At one extreme is Mildred Restall. "You see that vase over there?" she asks angrily, pointing to a white vase on her windowsill. "There's no such thing as the truth anymore. You can say that vase is black long enough until you believe it and it becomes the 'truth.' That's what I mean about Oak Island. Where did they get the idea that there's something down there? I ask you, *Where?*"

Mildred Restall has good reason to ask the question. While most Oak Island treasure hunters gave up their careers and life savings for the hunt, she gave up much more: the lives of her husband and firstborn son.

It happened on August 17, 1965. She and her husband, Bob, had been living on the island since 1959 while he hunted for the treasure. What happened that muggy day has never been entirely explained. Restall apparently was inspecting one of his pits when he blacked out and toppled in. His son Bobby came to his rescue, but when the other workers arrived they saw both father and son lying in the black water at the bottom. Four of them descended and were quickly overcome by fumes in the pit. Two were rescued, but the others, along with the Restalls, died—by drowning. The toxic gas was never identified.

Now Mildred lives alone in a bungalow; her living room looks across a patch of wild blueberries and chokecherries to the spruce-clad outline of Oak Island. She starts telling me about her early life, how she happened to end up on Oak Island.

"My husband and I," she says, "we rode the Globe of Death." On their motorcycles, Bob and Mildred would enter a sphere of steel mesh only sixteen feet in diameter. Then they would accelerate their bikes at right angles to each other, sometimes reaching speeds of nearly fifty miles an hour and crisscrossing each other's paths twice each revolution. Bob made vertical loops while Mildred roared around the globe's equator.

Mildred smiles and looks out to sea. "We had a good act—there's no getting away from it. You could just hear the gasp from the audience." After a moment she adds, almost to herself, "All the great circus acts were husband-and-wife teams."

She looks out her window, focusing on Oak Island in the distance, and sighs. "Why don't they just leave it go, let it stay a mystery?"

In 1968, a Montreal businessman named David Tobias took over in partnership with Dan Blankenship, who had been waiting for an opportunity to get involved. The following year they formed Triton Alliance Ltd.

Tobias attracted a strong group of investors, from the past president of the Toronto Stock Exchange to the chairman of one of Canada's largest chains of supermarkets. The company was initially capitalized at Can$520,000 and Tobias became its president.

Recently I visited Tobias in Montreal. Dressed in a blue flannel suit and smoking a pipe, he didn't look much like a treasure hunter. In fact, he says, he really isn't a treasure hunter.

Born in Winnipeg, Tobias grew up working while he attended high school and college at night. After World War II he became a salesman for a packaging company, and later acquired a company in the same industry. Today he lives in a large old house on the side of Mount Royal in Montreal. These accomplishments might satisfy most people, but not Tobias. "I wanted to do something," he says, "that no one had ever done before." That "something" is to solve the mystery of Oak Island once and for all. "Certainly I'm interested in finding something of value," he says, "but for me, the archaeological aspects are also important. We don't like to call ourselves a treasure hunt."

Will they be hiring a professional archaeologist?

"Yes, we surely plan to do that," Tobias says. The problem, he explains, is that so much sensationalism and hype have swirled around Oak Island that archaeologists have been scared away. "What this thing needed was a totally fresh approach," he says. "This is potentially a site with tremendous archaeological interest. This is important for Canada."

Triton has drilled more than two hundred cores on the island. The drills have gone much deeper than any of the shafts, right down into the bedrock at about 165 feet. Directly under the Money Pit, the drillers found a roughly circular hole in the bedrock that had been filled with puddled blue clay, earth that had been worked while wet to form an impervious mass—a waterproof plug. Inside the drill cores returned to the surface were bits of brass, charcoal, wood, china, and cement, along with oak buds. Still deeper the drillers found a natural horizontal cavity now filled with dirt. Cores from this region brought up more bits of wood and china.

"Then," Tobias recalls, "we went to Smith's Cove. I was hoping to make some dramatic breakthrough."

Several feet under the beach, they found a low rock wall, presumably a remnant of the box drains discovered a hundred years ago. Lying on top was a layer of fibrous material, which Tobias himself picked off and sent away for analysis. They also found the half-moon remains of an old wooden cofferdam across the mouth of the cove, as well as a curious heart-shaped stone, a pair of hand-wrought iron scissors, and other artifacts. The cofferdam remains were beyond where any previous dams had been built, suggesting that it was constructed by the original builders of the pit.

Tobias starts piling lab reports in front of me. Every artifact, in addition to the soil itself, has been exhaustively analyzed. Once again, the fibrous material was identified as coconut fiber, this time by the chief botanist at the National Museum of Natural Sciences in Ottawa. Several iron spikes from the cofferdam were analyzed by the Steel

Company of Canada and found probably to have been forged prior to 1790. The wood brought up from the Money Pit area dated to 1575, plus or minus eighty-five years. One sample was identified as crude lime cement, "likely to reflect human activity." The bits of brass and iron proved to be crude alloys whose microscopic structure suggested they had been made earlier than 1790.

Taken separately, the bits and pieces of evidence might not be all that conclusive. But taken together, they appear to add up to the fact that something happened on Oak Island at a great depth, prior to the 1795 discovery of the Money Pit.

To say that there are a number of theories is putting it mildly. Virtually every treasure reported missing over the past five hundred years has been placed, at one time or another, at the bottom of the Money Pit. The most enduring theory holds that Oak Island hides buried pirate treasure. The area was heavily frequented by pirates in the sixteenth and seventeenth centuries; Mahone Bay, where the island lies and a major feature of the Nova Scotia coastline, takes its name from the French word *mahonne*, a low-lying craft used by Mediterranean pirates.

The perennial favorite for this theory is Captain William Kidd. Kidd petitioned the House of Commons eleven days before his scheduled execution in 1701. He offered them a deal: if they would delay the hanging he would lead a fleet to the spot where he had buried his large East Indian treasure. His petition was refused and Kidd was executed on schedule. Unfortunately, there is no evidence that Kidd went anywhere near Nova Scotia.

Then there is the statement Blackbeard made shortly before his death and subsequent beheading. "I've buried my treasure," he bragged, "where none but Satan and myself can find it." If there ever was an accurate description of the Money Pit, this is it.

A more plausible theory maintains that Oak Island was a sort of Swiss bank for pirates. Similar caches have reportedly been found in Haiti and Madagascar, although such discoveries have not been confirmed by archaeologists. A number of pirate captains would band together, sink a deep central shaft, dig the appropriate number of side tunnels to accommodate the booty of each, then bore the flood tunnels that would leave the main shaft booby-trapped. At the site of one of these banks in Haiti, a number of heart-shaped stones similar to the one found at Smith's Cove were reportedly found.

Tobias thinks that Sir Francis Drake might have been responsible. In the late sixteenth century, Drake was given a secret commission from Queen Elizabeth I to prey on Spanish shipping, provided he turned the loot over to the Crown. Like other privateers, Drake may have decided to stash away part of the booty for himself, or the pit may have been part of a plan to have a treasure repository on this side of the Atlantic.

More recently, some historians have come to feel that no pirate crew could have had the discipline and organization to construct something as elaborate as the Money Pit. One historian who has taken more than a passing interest in Oak Island is Mendel Peterson, former chairman of the Department of Armed Forces History at the Smithsonian's Museum of American History. He headed the Institution's past programs in historical archaeology and underwater exploration. Peterson refused to speculate on who might have constructed the Money Pit. "But from what I know about the engineering involved, the resources that would have to have been mobilized, and the complicated structure of the Money Pit, I think it has to be the work of a government—or at least a very large, powerful organization. It couldn't have been pirates. Impossible."

Over the years, a veritable army of dowsers, mediums, soothsayers, automatic writers, spiritualists, psychics, card readers, channelers, and crank inventors have descended on Oak Island. They have located at the bottom of the pit everything from the secrets of the pyramids and the Holy Grail to the original drafts of Shakespeare's plays and the crown jewels of France.

OAK ISLAND ALSO has its skeptics, many of whom live in Western Shore, the town opposite Oak Island. For nearly two hundred years they've been watching treasure hunters come and go—empty-handed—

and hearing all the claims many times over. Clyde Vaughan, who is the great-great-great-grandson of Anthony Vaughan, one of the original discoverers of the Money Pit, thinks it's all a trifle ridiculous. "To tell you the truth, most people around here, they more or less laugh about the whole thing," he says placidly in his kitchen.

Some have suggested that the Money Pit might be nothing more than an old sinkhole. This, they claim, would explain the original depression in the ground, the log "platforms" (blowdowns that periodically washed into the pit) and the flood "traps" (natural watercourses in the bedrock). It would explain the hole in the bedrock found by Triton, and also how artifacts worked their way into deep caverns under the island. On the other hand, like most Oak Island theories, this one leaves a lot more unexplained than explained—notably the findings at Smith's Cove and the masses of coconut fiber. Mendel Peterson dismisses the sinkhole theory but thinks that the original engineers might have enlarged an existing sinkhole.

So how is Triton going to succeed where all others have failed? "Money," says Tobias. "Money and good planning."

The plans include the construction of an enormous shaft, the biggest yet, right in the Money Pit area. Eighty feet in diameter and extending about 200 feet into the ground, it will require the removal of at least 960,000 cubic feet of earth weighing close to 50,000 tons. The shaft was designed by Bill Cox of Cox Underground Research.

What about the water problem?

"All indications are that the water is at most a thousand gallons per minute," says Cox. "You hit a flow like that in a small shaft and, sure, it'll fill up awful quick. In a large shaft, you can stand there and watch it come in. Then you can pump it down and seal it off. We'll arm the site with six- to eight-thousand-gpm pumping capacity, ready to go into action immediately."

Most important, the shaft will be large enough to encompass most of the earlier workings and to place the walls in hard, virgin ground. As they go down, the interior walls will be carefully examined for signs of the postulated side tunnels leading to treasure.

Cox refuses to speculate about what might be down there. "My involvement," he says, "is to sink a shaft that will solve the mystery totally and forever, so completely that nobody will go back. It will be the decisive conclusion to Oak Island." For that reason the planned shaft has been nicknamed the "Decisive Conclusion" shaft.

Tobias himself hates to speculate on how much treasure might be down there, but at one point I did get a hint. We were having lunch in an elegant Montreal restaurant and I asked him once again.

He leaned forward. "There are some who say there could be as much as several billion down there."

If that's true, I replied, then this could be as big as the discovery of King Tut's tomb.

"I hate saying things like that," Tobias responded, "but, yes, it could be as great a discovery as King Tut's tomb."

Or it could be a $10 million hole.

UPDATE

A great deal has happened on Oak Island since the story was published. Triton Alliance was unable to launch its IPO on the Vancouver Stock Exchange due to the Black Monday stock market crash in October 1987. The Decisive Conclusion shaft was never sunk. In subsequent years, unable to raise adequate funds, the Triton partnership fell apart. In 2006, the brothers Rick and Marty Lagina of Michigan purchased David Tobias's share in the Oak Island property and with Dan Blankenship formed a new company to try to recover the treasure. The Lagina brothers have not found any evidence of treasure, but they struck it rich when they partnered with the History Channel on a reality television show, The Curse of Oak Island, *which began airing in 2014 and is still running. The mystery of the Oak Island treasure—what it is and who buried it, or even if it exists at all—remains unsolved.*

THE MYSTERY OF SANDIA CAVE

ON MAY 6, 1940, *Time* reported news of an event that shook the world of archaeology: Frank Hibben, a thirty-year-old scholar with the University of New Mexico, in Albuquerque, had uncovered in a cave in New Mexico's Sandia Mountains evidence of the oldest human culture in the New World. At that time, the key to an archaeological site lay in its stratigraphy—the layering of artifacts with such things as bone, charcoal, and sediments. After five years of excavation, Hibben reported that the stratigraphy of Sandia Cave had proved "outstanding." The first layer consisted of the usual cave debris: dust, pack-rat middens, and bat guano. Below that, student diggers with his expedition had struck a hard stalagmitic crust. In a preliminary report, Hibben described this layer as resting "absolutely unbroken over the whole of the cave floor." It was a terrific stroke of luck—as if, thousands of years earlier, the entire site had been entombed in concrete. As Hibben wrote in his 1946 book *The Lost Americans*, when the student diggers laid open the crust with sledgehammers and pickaxes "it was like opening the lid of a gigantic sardine can, whose contents would give you all the evidence you wanted for the earliest history of the New World." Underneath the crust they found a layer of artifacts mingled with the bones of extinct animals, he recounted, and these artifacts belonged to the Folsom people, who thrived in the Southwest perhaps ten thousand years ago and hunted bison with a spear point that had

---

Originally published in the *New Yorker* in 1995.

a unique, "fluted" shape. The Folsom people, whose culture had been discovered with great fanfare in the 1920s, were the earliest known human beings in the New World.

But Hibben and his colleagues kept digging. Below the Folsom level, they struck a smooth layer of yellow ochre, devoid of artifacts. The ochre was a fine-grained sedimentary deposit, laid down during a period of wetness in the cave, whose exact origin was mysterious. What lay beneath this layer was fabulous: the remains of an entirely unknown culture, thousands of years older than Folsom. Hibben theorized that the Sandia Cave people, as he named them, lived at least twenty-five thousand years ago and were big-game hunters. The bones of their prey—mammoths, mastodons, bison—lay scattered about, mixed with spear points, scrapers, and other tools. And the spear points were peculiar. They had a "shoulder," or notch, flaked into one side—presumably to provide a place to haft the point to the spear's shaft—and they looked like nothing that had been found in the Southwest before. Indeed, Hibben noted, these Sandia points bore an extraordinary resemblance to a European type of spear point, called the Solutrean, that had been found in France and Spain.

An archaeologist could hardly have discovered a better site: the only thing missing was the bones of Sandia man himself. In fact, it was almost too good to be true, for the archaeological community had been eagerly hoping for two discoveries—a pre-Folsom people and a connection to the Old World—and Sandia Cave provided both. The Sandia point fitted right in between the Old World Solutrean and the New World Folsom: it had a single shoulder, like the Solutrean, but some of the Sandia points also showed the fluting characteristic of Folsom—a technique known only in the New World. "It was the missing link that everyone had been looking for," one archaeologist told me. It implied that Sandia man, the earliest American, had a European origin.

Sandia man had a revolutionary impact on New World archaeology. Textbooks were revised. By the end of the 1940s, every student of American archaeology learned first about Sandia man—or the Sandia complex, as the culture was termed. In 1961, Congress declared

Sandia Cave a National Historic Landmark, and a bronze plaque was affixed below it, noting that the site possessed "national significance in commemorating the history of the United States of America." Today, Sandia Cave is a tourist site, with a parking lot, paths, signs, railings, and a spiral staircase leading to the cave itself. After all, if Hibben's twenty-five-thousand-year-old dating is correct, Sandia man represents a culture twice as old as any other generally accepted human culture in the New World.

The University of New Mexico, which was already respected for its archaeological work, became renowned, and Hibben himself became one of the country's most celebrated archaeologists. He was no slope-shouldered professor. Photographs of the young archaeologist show a strikingly handsome man with a square, rough-hewn face, brilliant blue eyes, and a Hemingway physique. He had a deep, vibrant voice, which thrilled the lecture hall, and a wicked sense of humor. Many say he was the most mesmerizing lecturer in the university's history. During his years of teaching—he formally retired in 1975, but continued to lecture, carry on research, and excavate—he probably inspired more young students to become archaeologists than anyone else in the country.

One former student, Dave Snow, recalled that when Hibben taught hominid evolution he would start the lecture by "grabbing the door jamb and swinging into the classroom like an ape, and then leap up on his desk and make apelike sounds." Snow laughed at the memory. "It may be corny as hell, but if you're a student it makes you think and remember. When I saw the Indiana Jones movie, the first thing I thought was, Jesus, this is Frank Hibben." George A. Agogino, an archaeologist who worked with Hibben, remembered him as "the most charming person you're ever going to meet," and went on to say, "He was the only person out in the field who would go over to everyone in their individual sleeping bags and ask if they were perfectly comfortable before he went to sleep. He usually had a lot of venison that he had shot, and he was a great cook. He used to have these things for breakfast he called black-eyed Susans, which consisted of a piece of toast with a hole cut out and an egg in the middle." Hibben loved to entertain students in his home as lavishly as if they were visiting dignitaries—something that was particularly well remembered and deeply appreciated by his Depression-bred students.

For sport, Hibben went big-game hunting in Africa and Asia, and he was considered one of this country's finest shots. He still regales people with stories of the chase. One of his favorites is of a lion that attacked him in Somalia; he killed it and then rode eighty miles by camel to Mogadishu to get medical help for his broken back. Often, Hibben ends the story by pointing to a trophy on the wall and saying, "That's the one over there, with the smile on his face." He shows guests a gigantic, man-eating leopard in his sitting room—allegedly the world-record specimen—and tells the harrowing story of how he stalked and killed it. He reminisces about hunting ibex and leopard with the Shah of Iran, shooting argali in the Altai Mountains of Outer Mongolia, and tracking mountain lions in the backcountry of New Mexico with Ben Lilly, the last of the mountain men.

According to his colleagues, Hibben would sometimes disappear for a week or two and come back looking jet-lagged and dropping hints about secret missions for the State Department or private meetings with such people as Henry Kissinger. When he took students to archaeological sites in New Mexico, he sometimes wore safari gear, enhanced with a leopard-skin hat and epaulets, and carried a large gun.

He was not as well liked among some of his colleagues as he was by his students, but one reason may have been jealousy. Most archaeologists could barely meet the rent on their chipboard bungalows; Hibben, who had married into money, lived in a large adobe house on the edge of campus, attended by servants. While other archaeologists lectured at ladies' teas and garden clubs, Hibben hobnobbed with oil millionaires, senators, and governors. Most archaeologists could barely write a literate sentence; Hibben churned out eight books and numerous spirited articles, some of which became classics in the field. He made charming and witty appearances on numerous television shows (including *What's My Line?* and *To Tell the Truth*), and was featured in a twenty-eight-part ABC television series called *On Safari with Frank*

Hibben. Most archaeologists labor for years at obscure sites; Hibben always seemed to strike archaeological gold.

Sandia Cave was only the first of several brilliant discoveries. Soon after, Hibben launched an expedition to Alaska seeking proof that Folsom man had migrated from Asia. In *The Lost Americans*, he told of braving vicious storms and high seas to land his boat in a remote inlet known as Chinitna Bay, and went on to reveal that there, below a smoking volcano, he had made a stunning discovery. "The glitter of flint among the dull pebbles in the sand caught our eyes," he wrote. "There was a familiar shape among the litter of material at the bottom of the bank. It was as though we had never left New Mexico at all. There was no doubt about it. Lying face up, with its characteristic groove and outline revealed at a glance, was a Folsom point!... We had back-tracked Folsom man almost to his starting point."

As Hibben and his team ranged up and down the shoreline, they found "everywhere" flint chips and bits of burned charcoal. But that was not all. "Protruding here and there from the bank, or shattered in sodden fragments on the beach, were the bones of mammoth. Mammoth gave us the time; the flint points gave us the picture. Here was a camp site of ancient men who had killed and eaten now-extinct animals." They made a quick survey before violent weather drove them from the bay.

"What we had found in a whole summer's cruise to Alaska could be contained in an old hat," Hibben wrote. "The implications involved, however, were epic-making. We had demonstrated that man came to the New World by the front door, across the Bering Strait, and had lived first in Alaska.... The first American was no longer a mystery." There had been, Hibben theorized, two early waves of human migration to the New World—Sandia was the first and Folsom the second—and both had ended up living in Sandia Cave.

When I visited Professor Hibben one morning last year [1994], I found him, at eighty-three, quite active. He had recently gone on safari to Africa, and he was still lecturing, researching, and writing. He seemed to be basking in a kind of autumnal glory. He had arranged for his house to go to the university on his death, to become an anthropological-research center, with a multimillion-dollar endowment. Richard Peck, the university's president, had told me that to his knowledge it was the largest private gift in the school's history. "Frank made our Anthropology Department renowned. His generous gift means that people will appreciate his work for a long time to come. It's got to be especially satisfying that his work has received a high academic reputation but also has a financial worth that doesn't often go along with scholarship."

When Professor Hibben greeted me, he was wearing khaki shorts, oxfords, and knee-high socks, and looked like a British explorer dressed for tiffin. Many of his colleagues had described to me his "presence," but I was not prepared to meet an eighty-three-year-old man who radiated such granitic strength, charm, and intellectual brilliance. I could understand why rooms fell silent when he entered, why he still packed the biggest lecture halls on campus, why so many were still afraid of him. He was the grand professor in the prewar style.

Meeting him should have been like meeting a childhood hero. When I was fourteen, I spent a summer in New Mexico on an archaeological dig. We trenched into a prehistoric trash heap near Gallup and uncovered four Anasazi skulls from around 1000 A.D. and some pots—an experience that forever fixed in me a passion for archaeology. All that summer, I carried around in my backpack a dog-eared copy of H. M. Wormington's *Prehistoric Indians of the Southwest*, the definitive work on the subject. I could recite passages almost by heart: "The earliest culture of the Western Hemisphere, about which we have any information, is the *Sandia*." The Old World may have had its Peking man, Neanderthal man, and Cro-Magnon man, but *we* had Sandia man.

Sadly, what had brought me to Professor Hibben's house was not a childlike hero worship but a journalist's obligation to record his responses to a number of serious allegations. Not long before, the subject of Sandia man had come up in a conversation I had with Tim Maxwell, the director of the office of archaeological studies for the Museum of New Mexico. To my astonishment, Maxwell had cast doubt on the discovery. As he has said, "There have been a lot of rumors and allegations about Sandia man. It's not just that they question the interpretation. They question the—damn, how should I phrase this?—the *authenticity* of the archaeological context."

SIXTY YEARS HAVE passed since the first Sandia point was unearthed in Sandia Cave, and many of the original excavators have died. I started a little archaeological excavation of my own—digging up the original reports, talking to retired archaeologists, peeling back the layers of history. And in the course of investigating various allegations about the site, I came to feel that I had stumbled into a vipers' nest. The Sandia Cave controversy has unfurled against a field—Southwestern archaeology—that is infamous for strife. In the entire discipline there may be no greater concentration of resentments and enmities. David Hurst Thomas, a curator at the American Museum of Natural History who has excavated major sites along the periphery of the Southwest, said, "I was always leery of getting into the mainstream there. It's common knowledge that Southwestern archaeologists eat their children." In interviews, I often heard archaeologists use such terms as "pathological liar," "mentally unbalanced," or "psychopath" to describe their colleagues. At other times, they would helpfully detail for me the sexual scandals, mental breakdowns, alcohol addictions, and job losses of their peers. The University of New Mexico's Department of Anthropology, as the leader in the field, was also the trendsetter for controversy.

I soon learned that questions had been raised about Sandia Cave from the beginning. In January 1940, a paper entitled "A Chronological Problem Presented by Sandia Cave, New Mexico" appeared in the journal *American Antiquity*. It had been written by Wesley L. Bliss, who had been a graduate assistant at the University of New Mexico in the late 1930s, and who worked on the earliest excavations in Sandia Cave with Frank Hibben.

In the article Bliss said that the layers in the cave had been mixed up by pack rats and other rodents—a stark contradiction of Hibben's assertion, in a preliminary 1937 report, that the stalagmitic crust was "absolutely unbroken" throughout the cave. This was no small quibble: a firm dating of Sandia man depended entirely on the layers' *not* having been disturbed. Bliss's statement threatened to discredit the discovery even before it was formally announced.

The University of New Mexico responded swiftly. The acting head of the Department of Anthropology, Donald D. Brand, wrote a letter to *American Antiquity* attacking Bliss's article and reasserting that the stalagmitic layer had been unbroken. In a reply to Brand's letter, which was also printed in *American Antiquity*, Bliss stubbornly held his ground. "I emphatically reaffirm my statement concerning *the rodent disturbances below the stalagmitic layer in the front portion of the cave*," he wrote, in italics, and he went on, "At the time, I called these sundry evidences of rodent disturbances to the attention of Dr. Frank C. Hibben."

Hibben himself responded this time, firing a tremendous barrage of words: "Unauthorized...flagrant...misleading...grossly and lamentably misinformed...unfortunate...inaccurate, premature, and unsubstantiated." Hibben also wrote, "The assumption that the writer [Hibben] was collaborating with this student [Bliss] in the excavating of this cave is so presumptuous as to be preposterous." While Hibben had not refuted a single point made in Bliss's article ("The glaring errors of this unfortunate article Professor Brand has already pointed out," he wrote), the response seemed to have the desired effect on everyone who read it. For a while, no more questions were raised about the cave.

I reached Bliss, who is eighty-nine, by telephone in Ojai, California, and learned that after Sandia he had gone on to a long career as a professor and archaeologist. An anger nearly sixty years old came crackling over our bad connection like electricity.

"I think he's a son of a bitch," Bliss said of Hibben. "He was a Harvard graduate, and Harvard graduates, in case you didn't know it, are supposed to be something special."

"Hibben claimed that it was an unbroken crust," I said to Bliss.

"Oh, he's *nuts*!" Bliss said.

Contrary to Hibben's assertions, Bliss had been intimately involved in excavating Sandia Cave and a nearby cave called Davis Cave. In the beginning, the excavation of Sandia Cave was informal—two eager young department assistants going to the mountains on weekends with some undergraduates to mess around in a cave. Exactly who was in charge, and what their relationship was, was never spelled out. If it had been thought an important site, it is doubtful that either Hibben or Bliss would have been in charge; a professor would have taken over.

Then, Bliss told me, at the close of the 1936 field season Hibben staged an archaeological coup. Bliss went up to the site, and Hibben simply told him to get out. Either Bliss did not protest very vigorously or the university's anthropology department sided with Hibben. The informality of the setup did Bliss in. Hibben also had justification for taking over Sandia. Initially, Bliss, as the more experienced digger, was given first choice of the two caves and selected the more promising-looking Davis Cave, which turned out to be sterile. Hibben worked on Sandia Cave. Bliss later worked in Sandia, but not as much as Hibben.

By the time Hibben took over the site, things had changed. He had become one of the most promising young archaeologists ever to attend the university. He had been accepted by Harvard into a doctoral program to study with the great physical anthropologist Earnest Hooton. He had married a wealthy socialite. Bliss was a "dirt" archaeologist, not a theoretician, and he did not show anywhere near the kind of promise that Hibben did, nor did he have connections or money. Although the first Sandia point was found when Bliss was still on the site, most of the important discoveries were made after Hibben took over.

FRANK CUMMINGS HIBBEN was born in 1910, in Lakewood, Ohio, to a respectable family of modest means. During the Depression, he was accepted at Harvard, but, for monetary reasons, went to Princeton, where his cousin John Grier Hibben was president. He graduated in 1933, and went to the University of New Mexico to take a master's degree in zoology; his thesis was a survey of mountain lions. In New Mexico, he was taken up by a wealthy older couple, Arthur and Eleanor (Brownie) Pack, who lavished attention on him. Arthur Pack ran *Nature* magazine and was deeply involved in conservation and wildlife research. (A forest in upstate New York is named after his father.) They also owned a vast spread in New Mexico known as Ghost Ranch—a place made famous by the paintings of Georgia O'Keeffe, a longtime ranch resident. Hibben lived with the Packs on the ranch and went with them on their travels.

Alden Hayes, a retired archaeologist who has known Hibben for about sixty years, explained to me what happened: "Arthur Pack grubstaked Frank. He financed Frank's work on mountain lions. While he"—Hibben—"was prowling around that part of the world, he kept stumbling over Indian ruins, and they kind of fascinated him. Arthur Pack's wife, Brownie, also fascinated him."

In his memoirs, Arthur Pack recalled how "every table in our house except the kitchen was spread with pottery shards." One day, he came home and found his wife gone, having run away with the student, who was twelve years her junior. She had left a note that, Pack wrote, "brought my world crashing down about me." She and Hibben were married on June 5, 1936, and to the marriage she brought a large amount of money and two stepdaughters. Hibben invested their money with remarkable shrewdness over the years and built up a considerable fortune, even as they lived extravagantly and traveled around the world. Today, his office contains well-thumbed stacks of *Barron's* and the *Wall Street Journal*.

"He spent enough time following those lions around with cowboys through the brush on horseback," Hayes recalled. "He picked up a wonderful repertoire of cowboy songs and the ability to tell a long, complicated, understated 'windy' with a Western drawl. He was a hell of a good storyteller, and a good fireside singer. He had an awful lot of charm. You could see why Brownie left old Arthur Pack."

STRATIGRAPHY WAS THE only way to date a Paleo-Indian site in 1940, but by the late forties a far more powerful method was being developed: carbon-14 dating. In 1951 or 1952, Hibben, in an effort to confirm the early date for Sandia, began shipping off to a new radiocarbon laboratory at the University of Michigan bone and ivory fragments that he identified as being from Sandia Cave. The early carbon-14 method was crude, time-consuming, and expensive. The carbon in the bone was turned into carbon dioxide gas, and then its faint natural radioactivity was measured with a Geiger counter. The first trial failed, because the radioactive cloud from an atomic bomb that the United States had detonated on Bikini Atoll passed over Michigan while the test was in progress. Reliable results finally came back on a sample of mammoth ivory. They showed a date of twenty-thousand-plus years, the extreme lower limit of the instrument. It appeared to be a stunning confirmation of the earlier, geological dating.

Once again, Sandia man made news across the country. *Time* ran another article. The *Times*, under the headline "MAN IN AMERICA '20,000 YEARS AGO,'" quoted Hibben as saying that there was "incontrovertible proof" that man existed in America that long ago.

Hibben published the results in the journal *Science* in 1955. In the article he said that the carbon dates merely confirmed others, which had been determined by an independent expert some years before. He wrote, "In 1948, two specimens of charcoal from fire hearths of the Sandia level of Sandia Cave were submitted by Kirk Bryan of Harvard University to ... the University of Chicago for C^{14} dating. ... From these two samples, tentative dates of 17,000-plus years ago and 20,000-plus years ago, respectively, were derived.

"Bryan was emphatic that these dates should not be published at the time, because of the inadequacy of the samples and the possibility of a considerable error."

A problem with this statement was that the University of Chicago radiocarbon laboratory, which was the only one in the world before 1951, had no records of any submissions from Kirk Bryan of Sandia Cave material. (Bryan died in 1950.) One of Bryan's colleagues wrote to *Science* demanding that Hibben's statement be retracted: "None of Bryan's intimate associates, including reputable archaeologists and geologists, some of whom were deeply concerned with the development of radiocarbon dating, can recall having heard Bryan mention the samples or the dates to which Hibben refers. Furthermore, Bryan's records have been searched and no reference to the alleged Sandia samples has been found."

Hibben beat a hasty retreat: he recast what had been an unequivocal statement into a vague misunderstanding. "In regard to the two samples of charcoal from Sandia Cave collected by the late Kirk Bryan," he responded, "there has been a dearth of evidence as to where and how Bryan dated these samples. I am in agreement . . . that these dates should be removed from the record." He added that this did not "invalidate the dates determined . . . from mammoth ivory from Sandia Cave."

But there *were* doubts about the mammoth ivory, which did not make the letters column of *Science*. Rumors began to circulate that the mammoth ivory had not come from Sandia Cave at all, and these rumors have persisted for forty years. In many conversations I had with archaeologists, I heard the same rumors, and I was able to track two of them to their sources.

The first allegation could be traced back to James Hester, the former state archaeologist of Colorado, now a professor emeritus at the University of Colorado. In 1952, Hester says, he was an undergraduate at the University of New Mexico, working as Hibben's assistant. "He was trying to get an early date on Sandia Cave," Hester told me, "and supposedly we were shipping specimens from Sandia Cave to a dating lab" at the University of Michigan. "At the same time, it seemed like he was doing things that were fraudulent."

Hester continued, "When I was working as his museum assistant, he told me that he wanted me to ship some tusk fragments he had in a cigar box." The next day, Hester said, "when I came back into the lab I saw that yellow ochre had been sprinkled all over them." Why someone might have done this was clear: a characteristic shared by all the Sandia specimens was that they were impregnated with this unusual yellow ochre. "Being at that time—what?—something like twenty years old, I decided the least controversial way to handle this was to just wash all the yellow ochre off and mail it," Hester said. "It seems to me that he got a date of about twenty thousand." This, then, was probably the tusk material whose dating was reported in the national press.

On another occasion, Hibben gave Hester some horse teeth to mail off for dating; Hibben said they were from Sandia Cave. As Hester sorted the teeth, he discovered a label on one with the word "Mousterian." The Mousterian people were a Middle Paleolithic culture of Eurasia, and if one was looking for a twenty-thousand-plus carbon-14 date Mousterian would do just fine. "I think I just didn't get around to mailing it," Hester said.

"How could Hibben have been so careless?"

"I have no idea," he said. "I was just an undergraduate, and I wasn't about to confront him with it."

"Do you believe that this was actual fraud?"

"That would be one fairly logical conclusion," he said dryly. "Or it may well be that a lot of what he did at Sandia Cave was simply not rigorous."

(Hibben, for his part, says he doesn't recall Hester's having worked for him.)

The second rumor I tracked to Lewis Binford, who is Distinguished Professor of Anthropology at Southern Methodist University. Binford, a brilliant theoretician, is considered to be one of the most influential archaeologists of his time. He is credited with being the father of "the new archaeology," an attempt to bring more rigorous scientific methods into the field. "He has inspired a lot of controversy, not only because of his revolutionary ideas but because of his very abrasive personality," one archaeologist told me. Binford and Hibben have a history of personal and professional disagreement.

Early in his career, Binford worked as assistant to a physicist in the C-14 laboratory at the University of Michigan. "Part of my job was to take care of all the samples and make sure all the information was correct," he told me. One of the samples that came in, he said, was a mammoth tooth, ostensibly from Sandia Cave. The tooth was studded with concretions that included pea gravel. In the report on Sandia Cave, Binford said, "there was... not a bit of mention of pea gravel. There was no way that mammoth tooth could have concretions full of pea gravel if there was no pea gravel in the deposits. So I wrote Frank Hibben a letter saying that we were trying to make sure there was no mix-up on what was being dated. And since he had sent in a lot of material—not just from Sandia Cave—was it possible that he had mislabeled the tooth." Binford received a letter from Hibben saying, in Binford's words, "that he was very sorry. That he *had* probably mixed up the samples and that I was correct—that the one with the pea gravel couldn't have come from Sandia Cave, and he'd gone back and was now sending specimens that were unequivocally from Sandia Cave." Enclosed was a box of mammoth-tusk fragments.

The mammoth-tusk fragments had a scaly crust of calcium carbonate. Calcium carbonate has an important quality: when it precipitates out of water, it often traps a unique blend of other minerals. If two specimens contain an identical blend, they probably came from the same place. (The tests are not conclusive.) Binford did a quick test on the mammoth-tusk fragments and the mammoth tooth with pea gravel that Hibben had originally sent.

What he found was puzzling. As Binford recalls it, Hibben's letter stated that the original mammoth tooth sent to the lab had *not* come from Sandia Cave. But Binford's tests revealed that the calcium carbonate on the tooth was the same as the calcium carbonate on the tusk fragments, which Hibben claimed *were* from Sandia Cave. In other words, the tests indicated that the tooth and the tusk fragments very probably came from the same place.

Binford wrote Hibben back, saying that he had tested the material and found these inconsistencies. "Hibben got angry and called my boss and said, 'I don't know who this person is and what the hell he's doing with my specimens.'"

In any event, they tested the tusk fragments and got the famous twenty-thousand-plus date. But that wasn't the end of the story. Hibben didn't want any of his specimens back. Binford kept the mammoth tooth with the pea gravel as a paperweight. Many years later, in 1968, he went to teach at the University of New Mexico. Two years after moving to Albuquerque, he began building a house. The drivers for the company supplying concrete for the job told him that they often found old bones in their gravel pits. "So one Sunday this truck driver came by and picked me up and we went to the gravel company's pits down in the Rio Grande Valley," Binford recalled. "What I saw immediately were these huge bands of this pea gravel that was exactly like what I had found on the mammoth tooth."

Binford said he collected some of the gravel from the pit and gave it and the mammoth tooth to a chemist at Los Alamos National Laboratory. A geological test was performed on both specimens. The pea gravel was, Binford recalled, the same.

To make a complicated story simple, Binford, using the calcium-carbonate test, determined that the mammoth tooth had probably come out of the gravel pit. (Hibben himself agrees that that is "almost certainly" the case.) Binford had also shown that the twenty-thousand-year-old "Sandia Cave" tusk fragments apparently had the same calcium-carbonate crust as the tooth. Therefore, those celebrated tusk fragments are likely to have come from the same place as the tooth: the gravel pit.

In his opinion, was Sandia Cave a fraud?

"Yes," Binford said. "But who it may have been in Frank's little network—which would, of course, include Frank—there's no way to say for sure.... There's no certainty that Frank did this, but it's hard to imagine that he excavated the site and he didn't know that these things weren't there.... Frank had a reputation in Albuquerque for—it would be kind to say—*stretching* things."

IN THE DECADE following the reported discovery at Sandia Cave, no other Sandia sites were found and no other Sandia points turned up in an archaeological context. (The word "context" is critical: a flint point taken out of context—that is, out of place—is worthless to archaeology and cannot be dated.) This was odd, since other important Paleo-Indian discoveries in the Southwest have normally been followed by a flood of corroborating finds. Then, in June 1954, Hibben announced that he had discovered another Sandia-man site.

It was situated on a remote ranch in the Estancia Valley of New Mexico, on the shores of a dry lake bed about fifty miles from Sandia Cave. Hibben had heard about the site from a man named Kenneth Kendall, who had been collecting flint points in the area for years. Hibben explored it with a graduate student named William Roosa. According to Roosa (whom I spoke with not long before his death), almost immediately upon their arrival they spied a mammoth bone sticking out of the sand. Hibben knelt down and quickly uncovered a beautiful Sandia point snuggled up against it. He turned to Roosa, saying, "Here's your dissertation."

Roosa excavated the site under Hibben's direction, and it yielded a number of Sandia points in association with mammoth bones. Several archaeologists, however, had their private doubts about the site; for one thing, Kendall, in all his years of collecting in the area, had never found a single-shouldered Sandia point.

The Lucy site, as it was called, after a nearby train station, augmented rumblings about Sandia man. Roosa told a few people at the time that he was suspicious about how easily Hibben had found the first Sandia point—that he was worried it had been planted. As Roosa uncovered more Sandia points, rumors began to fly that Hibben was planting points at the Lucy site for Roosa to find. When I spoke to Roosa, however, he said he eventually became convinced that the Sandia points at Lucy had *not* been planted. His dissertation treats the Sandia points as genuine artifacts.

IN THE EARLY 1960s, the Sandia rumors prompted two scientists to reexamine the Sandia Cave site in an attempt to verify Hibben's work. One was George Agogino, who is a professor emeritus in anthropology at Eastern New Mexico University, and who discovered the famous Hell Gap early-man site; the other was C. Vance Haynes Jr., Regents Professor in the Department of Anthropology and Geosciences at the

University of Arizona, who is considered to be the finest Paleo-Indian geochronologist in the country. "If you think you've got early man, Haynes is the man to convince," one archaeologist told me.

Haynes said to me, "Frankly, when I went into Sandia Cave, I had no question that everything was as told." But, he added, "I came out of that cave the first time we were up there and I said, 'George, I don't know what in hell's going on in that cave. There's *something* wrong here.'"

Their report, *Geochronology of Sandia Cave*, published in 1986 by the Smithsonian Institution as part of its Contributions to Anthropology series, said that the cave layers were very different from Hibben's description of them. This was not a stratigraphically "well-defined" site. Under the supposedly "continuous" stalagmitic crust and under the ochre layer, Haynes told me, they found a "labyrinth" of recent rodent tunnels that were "chucky-jam-full of piñon nuts and acorns, pieces of corncobs... fragments of charcoal, pieces of old newspaper, cigarette butts, and matches.... There were even unmentionables in there—condoms and whatever." Some of this material, obviously, could have accumulated after Hibben's excavation, but carbon-14 dating showed that rodents had been digging in and mixing up the cave layers for "at least thirteen thousand five hundred years." (Pack rats, in particular, are known for fanatically shifting objects about in their vast nests.)

Haynes reported that uranium-series dating showed that the layer of yellow ochre had to be more than two hundred and twenty-five thousand years old. (He recently told me that newer tests have pushed the date back to three hundred thousand years.) Also, extensive testing of bones from the cave resulted in dates of less than fourteen thousand years. There was a good reason for this: the cave had been sealed and free of animal life from at least three hundred thousand to about fourteen thousand years ago. Therefore, the *entire* "Sandia level" lying under the yellow ochre—bones, tusk fragments, teeth, and artifacts—must have been transported from above. And Sandia man, if he existed, could be no more than about fourteen thousand years old.

These findings cast additional doubt on whether the twenty-thousand-plus-year-old mammoth-tusk fragments came from the cave. But they raise an even more curious question: With everything in the cave having been churned up for a staggering hundred and thirty-five centuries, how was it possible that *all* Sandia points—nineteen of them—were somehow carried by rodents to the bottom layer only? Surely pack rats do not distinguish between projectile points and other cave junk.

I posed the question to Haynes.

"Don't think we didn't ask ourselves that same question," he said, referring to himself and Agogino. "It's very, very strange." He said some archaeologists suggested that this was evidence that the Sandia points had been planted. Haynes explained that this was not his view. He felt there had been *some* kind of very old site in the cave, but what, exactly, we will never know.

Agogino and a graduate student, Dominique E. Stevens, looked into the published record on Sandia Cave. Their analysis, which was meticulous, revealed a number of gross inconsistencies—what they termed "a fog of literary contradiction and confusion." They found, for example, that "certain stratigraphic discrepancies exist" between Hibben's preliminary 1937 report and a final report he'd done in 1941. They found so many inconsistencies in the data indicating the sites of crucial artifacts that it was impossible to say precisely *where* they had been unearthed.

"I was a friend of Hibben's," George Agogino said. "He's not a sleazy little fraud. He's a person who could charm his way into anything. And he probably charmed himself into believing things that weren't really there. I'm sure he believed that Sandia Cave was twenty-five thousand years old."

IF THE GEOLOGICAL dating was incorrect, and the stratigraphy was in doubt, and the carbon-14 dates were questionable, what evidence is left that Sandia man existed at all? The distinctive single-shouldered Sandia points. *Somebody* made them.

I called Dr. Bruce Bradley, a senior research archaeologist at the Crow Canyon Archaeological Center, a leading research and educational organization in southern Colorado. Bradley is one of the top experts in the country on prehistoric lithics—tools made out of flaked stone. He told me he had examined the Sandia points, which are housed at the University of New Mexico's Maxwell Museum, in the mid-1980s. "Several things really impressed me," he said, measuring his words very carefully. "When you get a collection of stuff that's found together in an assemblage, there's sort of a special look that they have. They've all basically gone through the same weathering conditions, especially if they're in a place like a cave or kill site. Even if there are different flints and obsidians, you get this homogeneous *feel*. It's like they *go* together."

The various Sandia points did not look to Bradley as if they belonged together. He explained that some looked as if they had been lying on the surface for a long time, weathered and polished by windblown sand, while other pieces looked very "fresh," as if they had been long buried and protected from weathering. Also, they were made from many different kinds of materials. Strangest of all, they were manufactured by the use of radically different chipping techniques. Some showed fine pressure-flaking and others crude percussion flaking, and still others were fluted—yet all these techniques were used to create the same single-shouldered form.

Bradley had then looked at the Sandia points under magnification. "Two of them clearly showed modern alteration," he said. "One of them has the little shoulder that looked like it was made by grinding the edge with a grinding wheel. You could see these little facets that looked like somebody had taken it and put it on a Carborundum wheel. That was very suspicious. The second one I saw with what looked like modern modification on it had very abrupt flaking toward the shoulder, very steep. Under magnification, I looked at those flake marks and I saw little remnants of steel or something like that—little streaks. It's exactly the kind of effect you get if you take a nail to the edge of an arrowhead and reflake it.

"It's fairly clear that all the pieces were old pieces with the exception of the modern modifications on two of them." Bradley was at pains to point out, "I'm not saying anything about the collection except what I can say from what I saw. I don't want to imply anything.... This is really just impressionistic." But when I asked him for his personal opinion, he said, "I think it's a hoax. But this is total, pure speculation."

He went on, "They needed an assemblage of artifacts that looked as Solutrean as you can get, and I think the motivation was to try to bolster the theory that Clovis"—a slightly older culture than Folsom—"originated as Solutrean, and the thing that's most distinctive about Solutrean is the shouldered point. If you could prove that Clovis originated in southwestern Europe and came across the Atlantic—my goodness!" Then he added, referring to the possibility of a hoax, "I have no way of knowing who did it."

I hunted down other Paleo-Indian-lithics experts who had examined the points. None had published their findings—some, perhaps, for the simple reason that they were afraid of having to spend years in court defending a libel suit. Some people I talked to kept emphasizing that Sandia man and his points are still known from only two sites—Sandia Cave and Lucy—and that both sites were excavated by the same man.

"Have there been any other finds of Sandia projectile points over any geographical area?" asked James B. Griffin, a research associate in the Department of Anthropology at the Smithsonian Institution and an emeritus professor from the University of Michigan, who worked in the radiocarbon laboratory that dated the Sandia specimens. "If you answer that question, I think you've got the *final* answer on Sandia Cave." He has also noted, "Nowhere in the country, in the years of *intensive* archaeological work that has taken place, has anything comparable to the Sandia complex ever turned up."

These leading experts on Paleo-Indians and Southwestern archaeology who have examined the remaining Sandia Cave points were willing to go on the record regarding the authenticity of the Sandia assemblage:

Jim Judge, a professor of anthropology at Fort Lewis College, in Durango, Colorado, and a former director of the Chaco Project, a monumental research project done in Chaco Canyon, one of the most significant Anasazi sites in the Southwest: "Some are original artifacts, some are reworked artifacts."

Robert York, a Western archaeologist who has worked for twenty-five years for various federal agencies: "A couple of the points looked to me like reworked Agate Basin"—that is, much more recent points that were tampered with in modern times. But, he noted, "Others look like they may be authentic."

David Meltzer, a professor of anthropology at Southern Methodist University: "Boy, it's a strange bunch of critters in those boxes. It's hard to make sense out of that site.... I don't buy it."

Lewis Binford: "I felt [two of the points] were archaic points that had a shoulder put in them."

There are a lot of ambivalent feelings about suggesting that a colleague has salted a site, the most heinous of archaeological crimes. Dennis Stanford, the chairman of the Department of Anthropology at the Smithsonian Institution and a former student of Hibben's, typified this ambivalence: "There were some of those"—Sandia points—"in the cave that were not planted. In fact, I don't know that *any* were planted." Stanford accepts the antiquity of some of the material. "I think there's some early stuff there, because of the differential cave etching"—microscopic surface pitting caused by wet, acidic conditions commonly found in caves—"and they may be extremely old, because of the degree of acid weathering." But the context, he said, is less convincing. "I'm highly suspicious of the Sandia points as a type, because you don't see them anywhere else other than Lucy—but, of course, Hibben had his hand in that as well." Stanford said that he likes Hibben. "And I hate to denigrate and assassinate his character. But it's hard to trust his work, because there are so many inconsistencies."

Where do the inconsistencies come from?

"Sloppy work, maybe. I'm trying to figure out what to say to you. Well, I really wouldn't want you to print what I think of it."

What about the rest of the single-shouldered points? Are they real? There is little doubt that they are genuine prehistoric Indian points, but they might have been planted in the cave. If they were, even though they may be thousands of years old, they are just as fake as a point chipped out yesterday. Archaeology is like real estate: location is everything. (Archaeologists call this "provenience.") This is the allegation that has been raised about the Sandia layer—that perhaps genuine artifacts from elsewhere were put there. In fact, more than one archaeologist expressed the view that the Folsom layer in Sandia Cave may also be a "put-up job," in the sense that genuine Folsom points from elsewhere had been introduced into the layer above Sandia—the best way, in the days before carbon-14 dating, of proving the Sandia level was earlier.

Single-shouldered points are not unique. They have turned up over the years in many areas. Archaeologists do not call them Sandia points, because they clearly come from many different cultures and time periods. They are not like Folsom or Clovis points, which belong to a single culture. Those who have examined these odd single-shouldered points have found that most of them appear to be accidents: points that were broken and resharpened in antiquity, or points that had a knot of stone in them that resisted flaking, or broken double-shouldered points. That a whole slew of such points—all different from each other—should be found in a single layer in a cave, without a lot of other types of points, is curious, indeed.

Obviously, we can never know with certainty what went on in Sandia Cave twenty-five thousand years ago (or in the years since). It should be remembered that most of those who actually dug in the cave as students feel that the assemblage of points is genuine. These people include Hibben's nemesis, Wesley Bliss. While he questioned much that Hibben did in the cave, he felt that the points had really been there. One of the original excavators, Ernst Blumenthal, also no friend of Hibben's, told me, "I was there when those things were found. I don't believe for one minute this b.s. about Frank salting the cave." He said he thought that it would have been too difficult for Hibben to plant the artifacts.

Another archaeologist, Richard S. MacNeish, who has raised a storm of controversy himself by reporting evidence of New World human beings even older than Sandia man, also argues that the points are real. "I think Frank Hibben has been much maligned in this matter.... Some people get their reputations ahead by crawling over the dead bodies of the people they knife in the back."

WHEREVER ANCIENT PEOPLE camped, they spent time flaking and resharpening their flint tools. Just as a modern hunter cleans his rifle upon returning home, ancient hunters would attend to their points. If Sandia Cave were a typical Paleo-Indian campsite, one might expect to find hundreds of waste flakes. But at Sandia Cave only six provenienced waste flakes were found at the entire Folsom level and only seven at the Sandia level.

What's more, the sizes of the artifacts in Sandia Cave appear to confound the laws of statistics. All the sediments in the cave were screened through quarter-inch mesh. Normally, one would expect a smooth distribution of artifact sizes starting at just over a quarter inch and going on up. Yet almost nothing found in the cave—either at the Folsom or the Sandia level—is between a quarter inch and an inch in size.

One archaeologist has noted that the sixty-two provenienced artifacts that came out of the Folsom and Sandia levels are made of fifty-three different materials. Only three types of flint have so far been identified from this collection: one was local and two came from outcrops in Texas 260 and 360 miles away. This is unusual but not impossible; Paleo-Indians did carry and trade flint over long distances. Hibben himself now says, "To my eye, the Sandia Cave collection ... is a mixed bag, possibly due to the fact that the original occupants of Sandia Cave were itinerants and not residents. The collection would seem to indicate roving bands of small numbers of hunters."

North American Paleo-Indians, in general, did not seem to like living in caves, preferring to camp in the open. Cave sites are rare. And Sandia is a particularly unpleasant cave—a cramped, slanting hole in the rock, where you have to stoop and sometimes crawl on your belly in order to get around. That people lived in it might be understandable if it were the only cave in the area. But just a hundred yards from

Sandia lies the much more pleasant and livable Davis Cave. (The Sandia excavators themselves lived there during the summer season.) And Davis Cave, as Bliss found, was sterile. Hibben speculates that Sandia Cave could have been much larger in Pleistocene times, and says that Kirk Bryan, of Harvard, believed that the cliff face has receded several meters since that era, "carrying away much of the original cave and much of the contents with it." He adds, "Dr. Bryan envisioned the original site as a fairly roomy cavern of several meters' width and at least four or five meters' depth with easy access from the terrace below. The present cave was used as a depository for garbage and debris."

Many of the important artifacts from Sandia Cave, including eight of the nineteen Sandia points, are missing from the Maxwell Museum, where they were kept. (This may mean nothing at all; museums often misplace items over the years, and Hibben says that when he returned to New Mexico in 1946, after serving in the military during the war, "many of the specimens of the Sandia collection—notes, photographs, etc.—were missing or in disarray.")

Another oddity is that most of the points found in Sandia Cave were whole. For Paleo-Indian campsites, one archaeologist told me, the ratio is more like fifty fragments to one complete point. Tony Baker, a nonacademic specialist in prehistoric lithics who has assembled one of the finest scientific collections of points in the Southwest, said, "Any time anyone shows me a complete Paleo point, the hackles go up on my neck, and when they show me two complete Paleo points then I *know* something is wrong. A complete Paleo point is very, very rare."

Only one other site that Hibben discovered has been scrutinized as carefully as Sandia Cave. That was the Folsom campsite he reported in Chinitna Bay, in Alaska. In 1978, a team of geologists and archaeologists took a float plane into Chinitna Bay to investigate.

"We went in believing we were going to find a very old site," E. James Dixon, the chief archaeologist with the expedition, said to me. "We were *shocked*."

There was no site: no flint chips, no fire hearths, no mammoth bones, absolutely no evidence of a Paleo-Indian camp. Nor is it likely there could ever have been such a site. A meticulous analysis of the area, including extensive carbon-14 dating, proved the layers where Hibben had reported finding the mammoth bones to be only a few hundred years old. Indeed, that entire section of Chinitna Bay was no more than five thousand years old, while mammoths had become extinct in Alaska at least ten thousand years ago.

When Dixon asked Hibben if he might examine the mammoth bones, Hibben said they had been lost. The report of the paleontologist who identified the mammoth bones had also been lost. And the famous projectile point was missing.

"Some people said, 'Well, what did you expect?'" Dixon remarked to me.

Sandia man exists in a kind of limbo. In the past two decades, most references to the Sandia complex have quietly disappeared from many textbooks; others have footnoted it as controversial or of questionable age. "You can still find textbooks with Sandia at the bottom of the chart," Dennis Stanford said. "But it hardly ever gets mentioned any longer." Most professors simply stopped teaching it. Nevertheless, Sandia man lives on in the minds of countless scholars, educated amateurs, and archaeological enthusiasts. He is like a dinner guest who arrives unexpectedly, proves himself an embarrassment, and is politely ignored but will not go away. "I think right now it's the general mood of most archaeologists to just ignore it and move on," Stanford told me. "If there was nothing there and everything was just planted, then so what? If there was something there that was misrepresented, then that's a problem. But we'll never know, will we? Unless Hibben signs a confession." And he roared with laughter.

HIBBEN IS STILL a prominent and powerful figure in New Mexico. He is close friends with a former governor, and has even gone digging with him. Many top educators and state legislators took his courses at the University of New Mexico, and the local newspapers love him. He has served in several political positions, including commissioner of the New Mexico Game and Fish Department. Many still fear Hibben. A state employee who was involved in preventing Hibben from renewing an excavating permit got a call from the lieutenant governor.

When I visited Hibben in person, it was with some trepidation. His house, tucked behind an adobe wall on a quiet side street off the university campus, was large but unpretentious—on the outside. Then a servant opened the door to reveal an enormous bull-elephant head, which must have weighed half a ton, glaring from the far wall. Exotic trophies—from dik-dik heads, no bigger than a cat, to the heads of rhino, lion, hippo, and Cape buffalo—bristled from every surface. The prize leopard dominated one room, rising up with a great snarl from an outcrop of fake rocks, his pelt faded by the New Mexico sun. A brace of Masai lion spears leaned against a far wall. In a manner reminiscent of castles imported from Europe by American tycoons, some of the walls themselves were built with stone blocks from Anasazi ruins.

Frank Hibben walked slowly in to greet me, as solid and square as a prize fighter. We sat down at the dining-room table, and he called for his files.

I asked about the dozen or so argali heads mounted on the wall, and the conversation immediately turned to Roy Chapman Andrews, whose books on exploring in Mongolia we had both read in our youth.

"I traced some of Andrews' trail when I got to Outer Mongolia," Hibben said. "I had an assignment from the U.S. government to go into Lop Nor.... I was planting a device to monitor Chinese atomic tests at Lop Nor. I'm not sure this has been declassified. I hope so. My instructions were to get as close to Lop Nor as possible and leave it at a high point."

On the pretext of a hunting trip, he explained, he flew into the Soviet-dominated part of Mongolia, carrying the device with him. In Kobdo, he gave his official interpreter "the slip" and met up with a Chinese Mongol who knew the Altai Mountains. They outfitted an expedition and went into the Altais by Bactrian camel. Working their way through the mountains, the two secretly crossed the Sino-Soviet border and rode several hundred miles into China, getting the device as close as they dared to Lop Nor. "I put it on a butte and pulled up the antenna—a red light went on," Hibben said. "I was pretty glad to get rid of it." As they headed back to the Soviet border, he said, he and his guide were attacked by a Chinese patrol on camels. "We came out from behind this ridge.... They were four or five hundred yards from us, and they were cutting us off. They began to shoot, and we could see the bullets kicking up the dust.... They cut in behind us and stopped, and then my camel began to show signs of distress—they had shot him through the stomach.... He fell and his shoulder pinned my foot and hurt it.... I got on the baggage camel and so we straggled back and we traveled all that night."

After this entertaining digression, I brought the conversation around to Sandia Cave. What, I eventually asked, did he think of the Haynes-and-Agogino report?

"They were rank amateurs at the time," Hibben growled. "Both of them became experts—self-styled experts.... And what I've done is to simply wait until they find another site. In other words, why argue about this site? We excavated to the best of our ability and interpreted it the very best we could. Now, go and find your own site."

What about the doubts concerning the Sandia points?

"I don't think there's any substitute for doing it yourself. If you're there with your own trowel, your own respirators, digging in those levels, there's no substitute for it. When you're looking at that stuff right there, you can interpret it better than any armchair expert back in the library. And if the interpretation isn't correct we'll change it as we get new evidence.... I'm not going to defend myself. Let them find their own sites."

What about the Mousterian label that James Hester found on the horse tooth supposedly from Sandia?

"It could have been. We were dealing with European materials at the same time."

I told him the story of Lewis Binford and the mammoth tooth with the pea gravel.

"Lew Binford wasn't here at the time of Sandia Cave. I'm not trying to discredit Binford—and then, of course, I became very good friends with him. He was thrown out of Michigan, because he was a disturbing influence.... Then he went to Chicago and was thrown out of there, because he was a card-carrying Communist, which he certainly was. And after that we were going to hire him, and I went to all the lengths I could to prevent his being hired here, because I'm very patriotic.... He's a very tempestuous character.... I'm just surprised. I thought Lew had mellowed. Maybe he's still bitter because I tried to keep him out of here. I thought we'd sort of kissed and made up, but apparently not.... Because that's a vicious story."

(Binford, in fact, received his Ph.D. at Michigan and left for a job at the University of Chicago. As for Hibben's charge that Binford was thrown out of Chicago for being a Communist, Binford told me, "My left-wing views had nothing to do with leaving. That was not an issue at Chicago. Hibben is somewhere to the right of Attila the Hun—he was the kind of person who thought *Roosevelt* was a Communist. I was not a Communist." When I asked Binford if he had an axe to grind, he laughed. "I'm not out to get Hibben. I just always thought of him as this giant *curiosity*. It's my understanding that for many years Hibben held the record for doing the classwork for a Ph.D. at Harvard in the shortest period of time. So here was an undeniably brilliant man. I used to wonder, Why did he go through all these *charades* to make himself important?")

"There are archaeologists," I said at one point in my conversation with Hibben, "who imply that you committed fraud in Sandia Cave."

"Well, I was afraid all this would come up."

Later, he added, "I don't like that word 'fraud,' and if it appears with somebody connected with it, I'll retaliate. I'll reply legally."

He also said, categorically, "I've never falsified anything in my whole life. As a matter of fact, quite the contrary. I've always left a major part of anything I've excavated for somebody else to look at and interpret with modern methods, which I perhaps don't have."

HIBBEN EXPRESSED SOME surprise at the questions, and felt he had been ambushed. He also felt that it was unreasonable to expect him to answer questions on short notice about events that took place sixty years ago. I therefore drew up a list of questions and faxed them to Hibben for his considered reply.

He responded with a ten-page fax. The fax did not address some of the more serious issues, including the question of how all the Sandia points could have been found underneath the layer of yellow ochre, which has now been proved to be more than three hundred thousand years old. Hibben's most important point—and it was a substantial one—was that he was not alone in excavating the site. Two towering scientists of the time, Frank H. H. Roberts, a top archaeologist for the Smithsonian, and Kirk Bryan, the Harvard geologist, were intimately involved in the excavations. "Dr. Bryan," Hibben wrote, "was in attendance on a daily basis at the excavation of Sandia Cave and Dr. Roberts almost as often.... If Dr. Bryan or Dr. Roberts would have had any questions about the stratigraphy or the disposition of the artifacts, they had ample opportunity to correct the Sandia Report before it was published." If this were true—that Roberts and Bryan excavated the cave side by side with Hibben—it would put the Sandia Cave controversy in a very different light.

The record, however, contradicts Hibben's assertion. Bryan's final report on Sandia Cave, published in the same monograph with Hibben's, starts, "The present report is based on brief visits to Sandia Cave in 1939 and 1940." Bryan went on to say that the geological "facts" of the cave were "mostly obtained by Hibben's excavations and recorded by him." Vance Haynes researched Bryan's involvement with the cave and said that interviews with Bryan's widow and a perusal of the late geologist's notes showed that he probably paid only two visits to the cave during the entire five-year excavation.

I asked Ernst Blumenthal, who worked on the Sandia Cave excavation in 1939 and 1940, whether Frank Roberts had dug in the cave.

He answered, "Frank Roberts? Bullshit.... Frank is dead, so you can't ask *him*."

"How many times did he come to the cave in total?" I asked.

"My guess would be twice."

"And then Kirk Bryan, the geologist?"

"Kirk was there once or twice, but he was not digging.... Believe me, I lived in the cave just to the north of Sandia Cave, Davis Cave. I slept there, we ate there, we were there every day."

Daniel McKnight and Charles Lange, two student workers during the last summer of the excavation, don't recall seeing either man visit the site.

Thus, there was essentially no independent examination of either the archaeology or the geology of the cave by Roberts or Bryan. The diggers were almost exclusively students. Donald Brand, the man ultimately in charge of the work, was an anthropologist who specialized in geography.

In his fax, Hibben asserted, "I left intact approximately 30 percent of the entire fill along the north wall of the cave... for future study," but, he claimed, relic hunters destroyed those deposits during the Second World War.

While Agogino recalls a small, three-foot-long block of material some sixty feet back from the entrance, in the sterile part of the cave, other witnesses have dismissed Hibben's claim out of hand. Blumenthal called the assertion "bullshit." Hibben's published site maps show that the cave was cleaned out, wall to wall, floor to ceiling, for twenty-four meters from the entrance. Vance Haynes said that his meticulous inspection of the cave in the early 1960s turned up no evidence that such a strip had ever existed. McKnight and Lange are also quite positive that the cave was cleaned out going way back, and that almost nothing was left. "My impression was that we dug it out from wall to wall," Lange said.

In his fax, Hibben said, "As to James Hester working in my laboratory during the Sandia Cave years, I do not recall his being there.... At any event, I was in Africa working on Paleolithic remains there during the time that the Sandia material was sent to... the University of Michigan."

Concerning the issue of Wesley Bliss's involvement in Sandia Cave, Hibben wrote, "At no time was I a partner of his nor was he on my regular excavating crew. Bliss excavated at least once on his own and Dr. Brand... told him to cease and desist." Hibben said that the "excavation permit was issued in my name."

Again, the written record and the recollections of people show that these statements are not true. According to a 1942 article by Douglas Byers, who was asked by *American Antiquity* to investigate the controversy between Bliss and Hibben, the two men explored the cave together and filed a mining claim on October 18, 1935, in the names of Bliss, Hibben, and three others. (In those days, archaeologists often posted a mining claim to mark a site on public land.) Furthermore, Byers wrote, on November 7, 1935, Donald Brand "applied for a permit to conduct archaeological and paleontological excavations in the caves, stating that he was to be in general charge of the work and that Wesley L. Bliss was the individual in charge of field work." Byers further stated that "Mr. Bliss and Mr. Hibben were leaders of separate field parties.... During the academic year 1936–1937, actually from October to January, Mr. Bliss was in charge of the work at Sandia Cave." All these facts are supported by the recollections of those who were involved in the discovery and excavation of the cave, including Wesley Bliss himself. Everyone recalled that Bliss was intimately involved until Hibben "elbowed Bliss out," as Lange put it.

Hibben defended his work in Chinitna Bay by noting that "my examinations were superficial and from the surface only. We were based upon a boat offshore and the tides were running 30 feet or more." He also said that "the shore was eroding very rapidly at that time."

This doesn't explain where the mammoth bones, flint chips, fire hearths, or Folsom projectile point came from—or where they went. The geologist with Dixon noted that the shore was actually *rising* from the sea. With the use of Hibben's own photographs, Dixon's team was able to get within a few feet of the exact place where Hibben said he found the material and determined that "little erosion of this shoreline" had occurred. Everything, Dixon said, appeared to be intact and precisely as Hibben described it—except that there was no evidence whatever of Paleo-Indian occupation. In case they were at the wrong spot, they closely examined miles of adjacent shoreline.

Finally, Hibben defended the uniqueness of the Sandia site by saying that the Clovis and Folsom people were also known from more or less "unique" sites. This is wildly incorrect: there are some twenty-five or thirty documented Folsom sites and ten or so Clovis sites.

I LEAVE INTERSTATE 25 at the northern end of the Sandia Mountains. The road passes the old Spanish town of Placitas—a cluster of adobe houses and double-wides—and turns to dirt with a great lurch. Fat cottonwood trees crowd the shoulders, forming a green tunnel that echoes with the rippling of Las Huertas Creek. The sun shines through the leaves as if through stained glass, and they chatter softly in a rising breeze.

The parking lot is marked with a sign: "SANDIA CAVE MAN TRAIL." The trail itself dips into a copse of oaks tangled with poison ivy and then rises again, emerging along an arid slope dotted with snakeweed and piñon. To the northwest, a view opens across the Rio Grande. The cave lies ahead, a black hole in the cliff. I climb the spiral staircase up the precipice. The words "SICK" and "LOVE ME MO" are scratched into the ironwork.

An iron cage projects from the cave mouth. I enter the cave, crouching under the low ceiling. The walls are covered with obscenities and unintelligible gang graffiti. Cigarette butts are scattered about the cave floor, and a crushed Pepsi can lies in a pool of muddy water. About fifteen feet inside, the cave is bricked up. There is a faint tang of urine.

I turn around and look back through the cave mouth. It is like staring through a dead eye into the world. Across Las Huertas Canyon, the synclinal thrust of the Sandia Mountains rises to meet the blinding blue sky. The mountain slopes are studded with brilliant-green aspens and darker stands of spruce and fir. A red-tailed hawk moves into my field of view, teetering on an updraft. There is no sign of human life. The view has remained unchanged for millennia.

I wonder: Twenty-five thousand years ago, did someone really stare from this hole to see these same mountains, these same skies?

UPDATE

About a year after the article was published, I got a call from one of the archaeologists I had interviewed. He told me he'd been aghast to read a short article that Hibben had published in an obscure newsletter called Teocintli. *In the article, Hibben claimed to have sued me and the* New Yorker *for libel, and that he was involved in negotiations for a settlement. The archaeologist wanted to know how this had happened. What had I gotten wrong in the article?*

It wasn't true, I said. There was no lawsuit. Hibben's claim was a complete fabrication; he hadn't sued anyone—it was just another lie. I said, "Didn't you yourself tell me he was a pathological liar? So why do you believe him now?"

The archaeologist roared with laughter. "By God, I did tell you that! But that's the thing about Hibben—his lies are always so believable!"

Sometime later Hibben published another note in Teocintli *saying that the* New Yorker *had offered to settle the lawsuit for one dollar and an admission of guilt. I was amused, but the* New Yorker *attorney I had worked with, who had vetted the piece for libel, didn't think it was funny at all. She was furious. She said to me, "I think we should wait for the New Mexico three-year statute of limitations on libel to run out on your piece, and then we'll sue* him *for libel!" But the* New Yorker *never did sue. Hibben passed away in 2002, much eulogized by the University of New Mexico, which created the Hibben Center for Archaeological Research in his honor, partially financed by a donation of $3.5 million Hibben gave to the university shortly before his death.*

THE MYSTERY OF HELL CREEK

IF, ON A certain evening about sixty-six million years ago, you had stood somewhere in North America and looked up at the sky, you would have soon made out what appeared to be a star. If you watched for an hour or two, the star would have seemed to grow in brightness, although it barely moved. That's because it was not a star but an asteroid, and it was headed directly for Earth at about forty-five thousand miles an hour. Sixty hours later, the asteroid hit. The air in front was compressed and violently heated, and it blasted a hole through the atmosphere, generating a supersonic shock wave. The asteroid struck a shallow sea where the Yucatán peninsula is today. In that moment, the Cretaceous period ended and the Paleogene period began.

A few years ago, scientists at Los Alamos National Laboratory used what was then one of the world's most powerful computers, the so-called Q Machine, to model the effects of the impact. The result was a slow-motion, second-by-second false-color video of the event. Within two minutes of slamming into Earth, the asteroid, which was at least six miles wide, had gouged a crater about eighteen miles deep and lofted twenty-five trillion metric tons of debris into the atmosphere. Picture the splash of a pebble falling into pond water, but on a planetary scale. When Earth's crust rebounded, a peak higher than Mt. Everest briefly rose up. The energy released was more than that

---

Originally published in the *New Yorker* in 2019 as "The Day the Dinosaurs Died."

of a billion Hiroshima bombs, but the blast looked nothing like a nuclear explosion, with its signature mushroom cloud. Instead, the initial blowout formed a "rooster tail," a gigantic jet of molten material, which exited the atmosphere, some of it fanning out over North America. Much of the material was several times hotter than the surface of the sun, and it set fire to everything within a thousand miles. In addition, an inverted cone of liquefied, superheated rock rose, spread outward as countless red-hot blobs of glass, called tektites, and blanketed the Western Hemisphere.

Some of the ejecta escaped Earth's gravitational pull and went into irregular orbits around the sun. Over millions of years, bits of it found their way to other planets and moons in the solar system. Mars was eventually strewn with the debris—just as pieces of Mars, knocked aloft by ancient asteroid impacts, have been found on Earth. A 2013 study in the journal *Astrobiology* estimated that tens of thousands of pounds of impact rubble may have landed on Titan, a moon of Saturn, and on Europa and Callisto, which orbit Jupiter—three satellites that scientists believe may have promising habitats for life. Mathematical models indicate that at least some of this vagabond debris still harbored living microbes. The asteroid may have sown life throughout the solar system, even as it ravaged life on Earth.

The asteroid was vaporized on impact. Its substance, mingling with vaporized Earth rock, formed a fiery plume, which reached halfway to the moon before collapsing in a pillar of incandescent dust. Computer models suggest that the atmosphere within fifteen hundred miles of ground zero became red hot from the debris storm, triggering gigantic forest fires. As the Earth rotated, the airborne material converged at the opposite side of the planet, where it fell and set fire to the entire Indian subcontinent. Measurements of the layer of ash and soot that eventually coated the Earth indicate that fires consumed about seventy percent of the world's forests. Meanwhile, giant tsunamis resulting from the impact churned across the Gulf of Mexico, tearing up coastlines, sometimes peeling up hundreds of feet of rock, pushing debris inland and then sucking it back out into deep water, leaving jumbled deposits that oilmen sometimes encounter in the course of deep-sea drilling.

The damage had only begun. Scientists still debate many of the details, which are derived from the computer models, and from field studies of the debris layer, knowledge of extinction rates, fossils and microfossils, and many other clues. But the overall view is consistently grim. The dust and soot from the impact and the conflagrations prevented all sunlight from reaching the planet's surface for months. Photosynthesis all but stopped, killing most of the plant life, extinguishing the phytoplankton in the oceans, and causing the amount of oxygen in the atmosphere to plummet. After the fires died down, Earth plunged into a period of cold, perhaps even a deep freeze. Earth's two essential food chains, in the sea and on land, collapsed. About seventy-five percent of all species went extinct. More than 99.9999 percent of all living organisms on Earth died, and the carbon cycle came to a halt.

Earth itself became toxic. When the asteroid struck, it vaporized layers of limestone, releasing into the atmosphere a trillion tons of carbon dioxide, ten billion tons of methane, and a billion tons of carbon monoxide; all three are powerful greenhouse gases. The impact also vaporized anhydrite rock, which blasted ten trillion tons of sulfur compounds aloft. The sulfur combined with water to form sulfuric acid, which then fell as an acid rain that may have been potent enough to strip the leaves from any surviving plants and to leach the nutrients from the soil.

Today, the layer of debris, ash, and soot deposited by the asteroid strike is preserved in the Earth's sediment as a stripe of black about the thickness of a notebook. This is called the KT boundary, because it marks the dividing line between the Cretaceous period and the Tertiary period. (The Tertiary has been redefined as the Paleogene, but the term "KT" persists.) Mysteries abound above and below the KT layer. In the late Cretaceous, widespread volcanoes spewed vast quantities of gas and dust into the atmosphere, and the air contained far higher levels of carbon dioxide than the air that we breathe now. The climate was tropical, and the planet was perhaps entirely free of ice. Yet scientists know very little about the animals and plants that were living at the time, and as a result they have been searching for fossil deposits as close to the KT boundary as possible.

One of the central mysteries of paleontology is the so-called "three-meter problem." In a century and a half of assiduous searching, almost no dinosaur remains have been found in the layers three meters, or about nine feet, below the KT boundary, a depth representing many thousands of years. Consequently, numerous paleontologists have argued that the dinosaurs were on the way to extinction long before the asteroid struck, owing perhaps to the volcanic eruptions and climate change. Other scientists have countered that the three-meter problem merely reflects how hard it is to find fossils. Sooner or later, they've contended, a scientist will discover dinosaurs much closer to the moment of destruction.

Locked in the KT boundary are the answers to our questions about one of the most significant events in the history of life on the planet. If one looks at the Earth as a kind of living organism, as many biologists do, you could say that it was shot by a bullet and almost died. Deciphering what happened on the day of destruction is crucial not only to solving the three-meter problem but also to explaining our own genesis as a species.

ON AUGUST 5, 2013, I received an email from a graduate student named Robert DePalma. I had never met DePalma, but we had corresponded on paleontological matters for years, ever since he had read a novel I'd written that centered on the discovery of a fossilized Tyrannosaurus rex killed by the KT impact. "I have made an incredible and unprecedented discovery," he wrote me, from a truck stop in Bowman, North Dakota. "It is extremely confidential and only three others know of it at the moment, all of them close colleagues." He went on, "It is far more unique and far rarer than any simple dinosaur discovery. I would prefer not outlining the details via e-mail, if possible." He gave me his cell-phone number and a time to call.

I called, and he told me that he had discovered a site like the one I'd imagined in my novel, which contained, among other things, direct victims of the catastrophe. At first, I was skeptical. DePalma was a scientific nobody, a Ph.D. candidate at the University of Kansas, and he said that he had found the site with no institutional backing and no collaborators. I thought that he was likely exaggerating, or that he might even be crazy. (Paleontology has more than its share of unusual people.) But I was intrigued enough to get on a plane to North Dakota to see for myself.

DePalma's find was in the Hell Creek geological formation, which outcrops in parts of North Dakota, South Dakota, Montana, and Wyoming, and contains some of the most storied dinosaur beds in the world. At the time of the impact, the Hell Creek landscape consisted of steamy, subtropical lowlands and floodplains along the shores of an inland sea. The land teemed with life and the conditions were excellent for fossilization, with seasonal floods and meandering rivers that rapidly buried dead animals and plants.

Dinosaur hunters first discovered these rich fossil beds in the late nineteenth century. In 1902, Barnum Brown, a flamboyant dinosaur hunter who worked at the American Museum of Natural History, in New York, found the first Tyrannosaurus rex here, causing a worldwide sensation. One paleontologist estimated that in the Cretaceous period Hell Creek was so thick with T. rexes that they were like hyenas on the Serengeti. It was also home to Triceratops and duckbills.

The Hell Creek Formation spanned the Cretaceous and the Paleogene periods, and paleontologists had known for at least half a century that an extinction had occurred then, because dinosaurs were found below, but never above, the KT layer. This was true not only in Hell Creek but all over the world. For many years, scientists believed that the KT extinction was no great mystery: over millions of years, volcanism, climate change, and other events gradually killed off many forms of life. But, in the late 1970s, a young geologist named Walter Alvarez and his father, Luis Alvarez, a nuclear physicist, discovered that the KT layer was laced with unusually high amounts of the rare metal iridium, which, they hypothesized, was from the dusty remains of an asteroid impact. In an article in *Science*, published in 1980, they proposed that this impact was so large that it triggered the mass extinction, and that the KT layer was the debris from that event. Most paleontologists rejected the idea that a sudden, random encounter with space junk had drastically altered the evolution of life on Earth. But as the years passed the evidence mounted, until, in a 1991 paper, the smoking gun was announced: the discovery of an impact crater buried under thousands of feet of sediment in the Yucatán peninsula, of exactly the right age, and of the right size and geochemistry, to have caused a worldwide cataclysm. The crater and the asteroid were named Chicxulub, after a small Mayan town near the epicenter.

One of the authors of the 1991 paper, David Kring, was so frightened by what he learned of the impact's destructive nature that he became a leading voice in calling for a system to identify and neutralize threatening asteroids. "There's no uncertainty to this statement: the Earth will be hit by a Chicxulub-size asteroid again, unless we deflect it," he told me. "Even a three-hundred-meter rock would end world agriculture."

In 2010, forty-one researchers in many scientific disciplines announced, in a landmark *Science* article, that the issue should be considered settled: a huge asteroid impact caused the extinction. But opposition to the idea remains passionate. The main competing hypothesis is that the colossal "Deccan" volcanic eruptions, in what would become India, spewed enough sulfur and carbon dioxide into the atmosphere to cause a climatic shift. The eruptions, which began before the KT impact and continued after it, were among the biggest in Earth's history, lasting hundreds of thousands of years, and burying half a million square miles of the Earth's surface a mile deep in lava. The three-meter gap below the KT layer, proponents argued, was evidence that the mass extinction was well under way by the time of the asteroid strike.

IN 2004, DEPALMA, at the time a twenty-two-year-old paleontology undergraduate, began excavating a small site in the Hell Creek Formation. The site had once been a pond, and the deposit consisted of very thin layers of sediment. Normally, one geological layer might represent thousands or millions of years. But DePalma was able to show that each layer in the deposit had been laid down in a single big rainstorm. "We could see when there were buds on the trees," he told me. "We could see when the cypresses were dropping their needles in the fall. We could experience this in real time." Peering at the layers was like flipping through a paleo-history book that chronicled decades of ecology in its silty pages. DePalma's adviser, the late Larry Martin, urged him to find a similar site, but one that had layers closer to the KT boundary.

Today [2019], DePalma, now thirty-seven, is still working toward his Ph.D. He holds the unpaid position of curator of vertebrate paleontology at the Palm Beach Museum of Natural History, a nascent and struggling museum with no exhibition space. In 2012, while looking for a new pond deposit, he heard that a private collector had stumbled upon an unusual site on a cattle ranch near Bowman, North Dakota. (Much of the Hell Creek land is privately owned, and ranchers will sell digging rights to whoever will pay decent money, paleontologists and commercial fossil collectors alike.) The collector felt that the site, a three-foot-deep layer exposed at the surface, was a bust: it was packed with fish fossils, but they were so delicate that they crumbled into tiny flakes as soon as they met the air. The fish were encased in layers of damp, cracked mud and sand that had never solidified; it was so soft that it could be dug with a shovel or pulled apart by hand. In July 2012, the collector showed DePalma the site and told him that he was welcome to it.

"I was immediately very disappointed," DePalma told me. He was hoping for a site like the one he'd excavated earlier: an ancient pond with fine-grained, fossil-bearing layers that spanned many seasons and years. Instead, everything had been deposited in a single flood. But as DePalma poked around he saw potential. The flood had entombed everything immediately, so specimens were exquisitely preserved. He found many complete fish, which are rare in the Hell Creek Formation, and he figured that he could remove them intact if he worked with painstaking care. He agreed to pay the rancher a certain amount for each season that he worked there. (The specifics of the arrangement, as is standard practice in paleontology, are a closely guarded secret. The site is now under exclusive long-term lease.)

The following July, DePalma returned to do a preliminary excavation of the site. "Almost right away, I saw it was unusual," he told me. He began shoveling off the layers of soil above where he'd found the fish. This "overburden" is typically material that was deposited long after the specimen lived; there's little in it to interest a paleontologist, and it is usually discarded. But as soon as DePalma started digging he noticed grayish-white specks in the layers which looked like grains of sand but which, under a hand lens, proved to be tiny spheres and elongated droplets. "I think, Holy shit, these look like microtektites!" DePalma recalled. Microtektites are the blobs of glass that form when molten rock is blasted into the air by an asteroid impact and falls back to Earth in a solidifying drizzle. The site appeared to contain microtektites by the million.

As DePalma carefully excavated the upper layers, he began uncovering an extraordinary array of fossils, exceedingly delicate but marvelously well preserved. "There's amazing plant material in there, all interlaced and interlocked," he recalled. "There are logjams of wood, fish pressed against cypress-tree root bundles, tree trunks smeared with amber." Most fossils end up being squashed flat by the pressure of the overlying stone, but here everything was three-dimensional, including the fish, having been encased in sediment all at once, which acted as a support. "You see skin, you see dorsal fins literally sticking straight up in the sediments, species new to science," he said. As he dug, the momentousness of what he had come across slowly dawned on him. If the site was what he hoped, he had made the most important paleontological discovery of the new century.

DEPALMA GREW UP in Boca Raton, Florida, and as a child he was fascinated by bones and the stories they contained. His father, Robert Sr., practices endodontic surgery in nearby Delray Beach; his great-uncle Anthony, who died in 2005, at the age of a hundred, was a renowned orthopedic surgeon who wrote several standard textbooks on the subject. (Anthony's son, Robert's cousin, is the film director Brian De Palma.)

"Between the ages of three and four, I made a visual connection with the gracefulness of individual bones and how they fit together as a system," DePalma told me. "That really struck me. I went after whatever on the dinner table had bones in it." His family buried their dead pets in one spot and put the burial markers in another, so that he wouldn't dig up the corpses; he found them anyway. He froze dead lizards in ice-cube trays, which his mother would discover when she had friends over for iced tea. "I was never into sports," he said. "They tried to get me to do that so I would get along with the other kids. But I was digging up the baseball field looking for bones."

DePalma's great-uncle Anthony, who lived in Pompano Beach, took him under his wing. "I used to visit him every other weekend and show him my latest finds," DePalma said. When he was four, someone at a museum in Texas gave him a fragment of dinosaur bone, which he took to his great-uncle. "He taught me that all those little knobs and rough patches and protrusions on a bone had names, and that the bone also had a name," DePalma said. "I was captivated." At six or seven, on trips to Central Florida with his family, he started finding his own fossilized bones from mammals dating back to the Ice Age. He found his first dinosaur bone when he was nine, in Colorado.

In high school, during the summer and on weekends, DePalma collected fossils, made dinosaur models, and mounted skeletons for the Graves Museum of Archaeology and Natural History, in Dania Beach. He loaned the museum his childhood fossil collection for display, but in 2004 the museum went bankrupt and many of the specimens were carted off to a community college. DePalma had no paperwork to prove his ownership, and a court refused to return his fossils, which numbered in the hundreds. They were mostly locked away in storage, unavailable for public display and enjoyment.

Dismayed by what he called the "wasteful mismanagement" of his collection, DePalma adopted some unusual collecting practices. Typically, paleontologists cede the curation and the care of their specimens to the institutions that hold them. But DePalma insists on contractual clauses that give him oversight of the management of his specimens. He never digs on public land, because of what he considers excessive government red tape. But, without federal support for his work, he must cover almost all the costs himself. His out-of-pocket expenses for working the Hell Creek site amount to tens of thousands of dollars. He helps defray the expenses by mounting fossils, doing reconstructions, and casting and selling replicas for museums, private collectors, and other clients. At times, his parents have chipped in. "I squeak by," he said. "If it's a toss-up between getting more PaleoBond"—an expensive liquid glue used to hold fossils together—"or changing the air-conditioning filter, I'm getting the PaleoBond." He is single, and shares a three-bedroom apartment with casts of various dinosaurs, including one of a *Nanotyrannus*. "It's hard to have a life outside of my work," he said.

DePalma's control of his research collection is controversial. Fossils are a big business; wealthy collectors pay hundreds of thousands of dollars, even millions, for a rare specimen. (In 1997, a T. rex nicknamed Sue was sold at a Sotheby's auction, to the Field Museum of Natural History, in Chicago, for more than $8.3 million.) The American market is awash in fossils illegally smuggled out of China and Mongolia. But in the U.S. fossil collecting on private property is legal, as is the buying, selling, and exporting of fossils. Many scientists view this trade as a threat to paleontology and argue that important fossils belong in museums. "I'm not allowed to have a private collection of anything I'm studying," one prominent curator told me. DePalma insists that he maintains "the best of both worlds" for his fossils. He has deposited portions of his collection at several nonprofit institutions, including the University of Kansas, the Palm Beach Museum of Natural History, and Florida Atlantic University; some specimens are temporarily housed in various analytical labs that are conducting tests on them—all overseen by him.

In 2013, DePalma briefly made news with a paper he published in the *Proceedings of the National Academy of Sciences*. Four years earlier, in Hell Creek, he and a field assistant, Robert Feeney, found an odd, lumpy growth of fossilized bone that turned out to be two fused vertebrae from the tail of a hadrosaur, a duck-billed dinosaur from the Cretaceous period. DePalma thought that the bone might have grown around a foreign object and encased it. He took it to Lawrence Memorial Hospital, in Kansas, where a CT technician scanned it for free in the middle of the night, when the machine was idle. Inside the nodule was a broken tyrannosaur tooth; the hadrosaur had been bitten by a tyrannosaur and escaped.

The discovery helped refute an old hypothesis, revived by the formidable paleontologist Jack Horner, that T. rex was solely a scavenger. Horner argued that T. rex was too slow and lumbering, its arms too puny and its eyesight too poor, to prey on other creatures. When DePalma's find was picked up by the national media, Horner dismissed it as "speculation" and merely "one data point." He suggested an alternative scenario: the T. rex might have accidentally bitten the tail of a sleeping hadrosaur, thinking that it was dead, and then "backed away" when it realized its mistake. "I thought that was absolutely preposterous," DePalma told me. At the time, he told the *Los Angeles Times*, "A scavenger doesn't come across a food source and realize all of a sudden that it's alive." Horner eventually conceded that T. rex may have hunted live prey. But, when I asked Horner about DePalma recently, he said at first that he didn't remember him: "In the community, we don't get to know students very well."

Without his Ph.D., DePalma remains mostly invisible, awaiting the stamp of approval that signals the beginning of a serious research career. Several paleontologists I talked to had not heard of him. Another, who asked not to be named, said, "Finding that kind of fossil was pretty cool, but not life-changing. People sometimes think I'm dumb because I often say I don't have the answers—we weren't there when a fossil was formed. There are other people out there who say they do know, and he's one of those people. I think he can overinterpret."

AFTER RECEIVING DEPALMA'S email, I made arrangements to visit the Hell Creek site; three weeks later I was in Bowman. DePalma pulled up to my hotel in a Toyota 4Runner, its stereo blasting the theme to *Raiders of the Lost Ark*. He wore a coarse cotton work shirt, cargo pants with canvas suspenders, and a suede cowboy hat with the left brim snapped up. His face was tanned from long days in the sun and he had a five-day-old beard.

I got in, and we drove for an hour or so, turning through a ranch gate and following a maze of bone-rattling roads that eventually petered out in a grassy basin. The scattered badlands of Hell Creek form an otherworldly landscape. This is far-flung ranching and farming country; prairies and sunflower fields stretch to the horizon, domed by the great blue skies of the American West. Roads connect small towns—truck stop, church, motel, houses, and trailers—and lonely expanses roll by in between. Here and there in the countryside, abandoned farmhouses lean into the ground. Over millions of years, the Hell Creek layer has been heavily eroded, leaving only remnants, which jut from the prairie like so many rotten teeth. These lifeless buttes and pinnacles are striped in beige, chocolate, yellow, maroon, russet, gray, and white. Fossils, worked loose by wind and rain, spill down the sides.

When we arrived, DePalma's site lay open in front of us: a desolate hump of gray, cracked earth, about the size of two soccer fields. It looked as if a piece of the moon had dropped there. One side of the deposit was cut through by a sandy wash, or dry streambed; the other ended in a low escarpment. The dig was a three-foot-deep rectangular hole, sixty feet long by forty feet wide. A couple of two-by-fours, along with various digging tools and some metal pipe for taking core samples, leaned against the far side of the hole. As we strolled around the site, I noticed on DePalma's belt a long fixed-blade knife and a sheathed bayonet—a Second World War relic that his uncle gave him when he was twelve, he said.

He recalled the moment of discovery. The first fossil he removed, earlier that summer, was a five-foot-long freshwater paddlefish. Paddlefish still live today; they have a long bony snout, with which they probe murky water in search of food. When DePalma took out the fossil, he found underneath it a tooth from a mosasaur, a giant carnivorous marine reptile. He wondered how a freshwater fish and a marine reptile could have ended up in the same place, on a riverbank at least several miles inland from the nearest sea. (At the time, a shallow body of water, called the Western Interior Seaway, ran from the proto–Gulf of Mexico up through part of North America.) The next day, he found a two-foot-wide tail from another marine fish; it looked as if it had been violently ripped from the fish's body. "If the fish is dead for any length of time, those tails decay and fall apart," DePalma said. But this one was perfectly intact, "so I knew that it was transported at the time of death or around then." Like the mosasaur tooth, it had somehow ended up miles inland from the sea of its origin. "When I found that, I thought, There's no way, this can't be right," DePalma said. The discoveries hinted at an extraordinary conclusion that he wasn't quite ready to accept. "I was ninety-eight percent convinced at that point," he said.

The following day, DePalma noticed a small disturbance preserved in the sediment. About three inches in diameter, it appeared to be a crater formed by an object that had fallen from the sky and plunked down in mud. Similar formations, caused by hailstones hitting a muddy surface, had been found before in the fossil record. As DePalma shaved back the layers to make a cross section of the crater, he found the thing itself—not a hailstone but a small white sphere—at the bottom of the crater. It was a tektite, about three millimeters in diameter—the fallout from an ancient asteroid impact. As he continued excavating, he found another crater with a tektite at the bottom, and another, and another. Glass turns to clay over millions of years, and these tektites were now clay, but some still had glassy cores. The microtektites he had found earlier might have been carried there by water, but these had been trapped where they fell—on what, DePalma believed, must have been the very day of the disaster.

"When I saw that, I knew this wasn't just any flood deposit," DePalma said. "We weren't just near the KT boundary—this whole site *is* the KT boundary!" From surveying and mapping the layers,

DePalma hypothesized that a massive inland surge of water flooded a river valley and filled the low-lying area where we now stood, perhaps as a result of the KT-impact tsunami, which had roared across the proto-Gulf and up the Western Interior Seaway. As the water slowed and became slack, it deposited everything that had been caught up in its travels—the heaviest material first, up to whatever was floating on the surface. All of it was quickly entombed and preserved in the muck: dying and dead creatures, both marine and freshwater; plants, seeds, tree trunks, roots, cones, pine needles, flowers, and pollen; shells, bones, teeth, and eggs; tektites, shocked minerals, tiny diamonds, iridium-laden dust, ash, charcoal, and amber-smeared wood. As the sediments settled, blobs of glass rained into the mud, the largest first, then finer and finer bits, until grains sifted down like snow.

"We have the whole KT event preserved in these sediments," DePalma said. "With this deposit, we can chart what happened the day the Cretaceous died." No paleontological site remotely like it had ever been found, and, if DePalma's hypothesis proves correct, the scientific value of the site will be immense. When Walter Alvarez visited the dig last summer, he was astounded. "It is truly a magnificent site," he wrote to me, adding that it's "surely one of the best sites ever found for telling just what happened on the day of the impact."

WHEN DEPALMA FINISHED showing me the dig, he introduced me to a field assistant, Rudy Pascucci, the director of the Palm Beach Museum. Pascucci, a muscular man in his fifties, was sunburned and unshaven, and wore a sleeveless T-shirt, snakeproof camouflage boots, and a dusty Tilley hat. The two men gathered their tools, got down on the floor of the hole, and began probing the three-foot-high walls of the deposit.

For rough digging, DePalma likes to use his bayonet and a handheld Marsh pick, popularized by the nineteenth-century Yale paleontologist Othniel C. Marsh, who pioneered dinosaur hunting in the American West and discovered eighty new species. The pick was given to him by David Burnham, his thesis adviser at Kansas, when he completed his master's degree. For fine work, DePalma uses X-Acto knives and brushes—the typical tools of a paleontologist—as well as dental instruments given to him by his father.

The deposit consisted of dozens of thin layers of mud and sand. Lower down, it graded into a more turbulent band of sand and gravel, which contained the heavier fish fossils, bones, and bigger tektites. Below that layer was a hard surface of sandstone, the original Cretaceous bedrock of the site, much of which had been scoured smooth by the flood.

Paleontology is maddening work, its progress typically measured in millimeters. As I watched, DePalma and Pascucci lay on their stomachs under the beating sun, their eyes inches from the dirt wall, and picked away. DePalma poked the tip of an X-Acto into the thin laminations of sediment and loosened one dime-size flake at a time; he'd examine it closely, and, if he saw nothing, flick it away. When the chips accumulated, he gathered them into small piles with a paintbrush; when those piles accumulated, Pascucci swept them into larger piles with a broom and then shoveled them into a heap at the far end of the dig.

Occasionally, DePalma came across small plant fossils—flower petals, leaves, seeds, pine needles, and bits of bark. Many of these were mere impressions in the mud, which would crack and peel as soon as they were exposed to the air. He quickly squirted them with Paleo-Bond, which soaked into the fossils and held them together. Or, using another technique, he mixed a batch of plaster and poured it on the specimen before it fell apart. This would preserve, in plaster, a reverse image of the fossil; the original was too short-lived to be saved.

When the mosquitoes got bad, DePalma took out a briar pipe and packed it with Royal Cherry Cavendish tobacco. He put a lighter to it and vigorously puffed, wreathing himself in sickly-sweet smoke, then went back to work. "I'm like a shopaholic in a shoe store," he said. "I want everything!"

He showed me the impression of a round object about two inches wide. "This is either a flower or an echinoderm," he said, referring to a group of marine life-forms that includes sea urchins and starfish.

"I'll figure it out in the lab." He swiftly entombed it in PaleoBond and plaster. Next, he found a perfect leaf, and near that a seed from a pinecone. "Cretaceous mulch," he said, dismissively; he already had many similar examples. He found three more small craters with tektites in them, which he sectioned and photographed. Then his X-Acto blade turned up a tiny brown bone—a jaw, less than a quarter inch in length. He held it up between his fingers and peered at it with a lens.

"A mammal," he said. "This one was already dead when it was buried." Weeks later, in the lab, he identified the jaw as probably belonging to a mammal distantly related to primates—including us.

Half an hour later, DePalma discovered a large feather. "Every day is Christmas out here," he said. He exposed the feather with precise movements. It was a crisp impression in the layer of mud, perhaps thirteen inches long. "This is my ninth feather," he said. "The first fossil feathers ever found at Hell Creek. I'm convinced these are dinosaur feathers. I don't know for sure. But these are primitive feathers, and most are a foot long. There are zero birds that big from Hell Creek with feathers this primitive. It's more parsimonious to suggest it was a known dinosaur, most likely a theropod, possibly a raptor." He kept digging. "Maybe we'll find the raptor that these feathers came from, but I doubt it. These feathers could have floated from a long way off."

His X-Acto knife unearthed the edge of a fossilized fin. Another paddlefish came to light; it later proved to be nearly six feet long. DePalma probed the sediment around it, to gauge its position and how best to extract it. As more of it was exposed, we could clearly see that the fish's two-foot-long snout had broken when it was forced—probably by the flood's surge—against the branches of a submerged araucaria tree. He noted that every fish he'd found in the site had died with its mouth open, which may indicate that the fish had been gasping as they suffocated in the sediment-laden water.

"Most died in a vertical position in the sediment, didn't even tip over on their sides," he said. "And they weren't scavenged, because whatever would have dug them up afterward was probably gone." He chipped away around the paddlefish, exposing a fin bone, then a half-dollar-size patch of fossilized skin with the scales perfectly visible. He treated these by saturating them with his own special blend of hardener. Because of the extreme fragility of the fossils, he would take them back to his lab, in Florida, totally encased in sediment, or "matrix." In the lab, he would free each fossil under a magnifying glass, in precisely controlled conditions, away from the damaging effects of sun, wind, and aridity.

As DePalma worked around the paddlefish, more of the araucaria branch came to light, including its short, spiky needles. "This tree was alive when it was buried," he said. Then he noticed a golden blob of amber stuck to the branch. Amber is preserved tree resin and often contains traces of whatever was in the air at the time, trapping the atmospheric chemistry and even, sometimes, insects and small reptiles. "This is Cretaceous flypaper," he said. "I can't wait to get this back to the lab."

An hour later, he had chiseled all the way around the fish, leaving it encased in matrix, supported by a four-inch-tall pedestal of rock. "I'm pretty sure this is a species new to science," he said. Because the soft tissue had also fossilized, he said, even the animal's stomach contents might still be present.

He straightened up. "Time to plaster," he said. He took off his shirt and began mixing a five-gallon bucket of plaster with his hands, while Pascucci tore strips of burlap. DePalma took a two-by-four and sawed off two foot-long pieces and placed them like splints on either side of the sediment-encased fossil. One by one, he dipped the burlap strips in the plaster and draped them across the top and the sides of the specimen. He added rope handles and plastered them in. An hour later, when the plaster had cured, he chiseled through the rock pedestal beneath the fossil and flipped the specimen over, leaving the underside exposed. Back in the lab, he would go through this surface to access the fossil, with the plaster jacket acting as a cradle below. Using the rope handles, DePalma and Pascucci lugged the specimen, which weighed perhaps two hundred pounds, to the truck and loaded it into the back. Later, DePalma would store it behind a friend's ranch house, where all his jacketed fossils from the season were laid out in rows, covered with tarps.

DePalma resumed digging. Gusts of wind stirred up clouds of dust, and rain fell; when the weather cleared, the late-afternoon sun spilled across the prairie. DePalma was lost in another day, in another time. "Here's a piece of wood with bark-beetle traces," he said. Plant fossils from the first several million years after the impact show almost no signs of such damage; the insects were mostly gone. The asteroid had likely struck in the fall, DePalma speculated. He had reached this conclusion by comparing the juvenile paddlefish and sturgeon he'd found with the species' known growth rates and hatching seasons; he'd also found the seeds of conifers, figs, and certain flowers. "When we analyze the pollen and diatomaceous particles, that will narrow it down," he said.

In the week that followed, fresh riches emerged: more feathers, leaves, seeds, and amber, along with several other fish, three to five feet long, and a dozen more craters with tektites. I have visited many paleontological sites, but I had never seen so many specimens found so quickly. Most digs are boring; days or weeks may pass with little found. DePalma seemed to make a noteworthy discovery about every half hour.

When DePalma first visited the site, he noted, partially embedded on the surface, the hip bone of a dinosaur in the ceratopsian family, of which Triceratops is the best-known member. A commercial collector had tried to remove it years earlier; it had been abandoned in place and was crumbling from years of exposure. DePalma initially dismissed it as "trash" and decried the irresponsibility of the collector. Later, though, he wondered how the bone, which was heavy, had arrived there, very close to the high-water mark of the flood. It must have floated, he said, and to have done so it must have been encased in desiccated tissue—suggesting that at least one dinosaur species was alive at the time of the impact. He later found a suitcase-size piece of fossilized skin from a ceratopsian attached to the hip bone.

At one point, DePalma set off to photograph the layers of the deposit that had been cut through and exposed by the sandy wash. He scraped smooth a vertical section and misted it with water from a spray bottle to bring out the color. The bottom layer was jumbled; the first rush of water had ripped up layers of mud, gravel, and rocks and tumbled them about with pieces of burned (and burning) wood.

Then DePalma came to a faint jug-shaped outline in the wall of the wash. He examined it closely. It started as a tunnel at the top of the KT layer, went down, and then widened into a round cavity, filled with soil of a different color, which stopped at the hard sandstone of the undisturbed bedrock layer below. It looked as though a small animal had dug through the mud to create a hideout. "Is that a burrow?" I asked.

DePalma scraped the area smooth with his bayonet, then sprayed it. "You're darn right it is," he said. "And this isn't the burrow of a small dinosaur. It's a mammal burrow." (Burrows have characteristic shapes, depending on the species that inhabit them.) He peered at it, his eyes inches from the rock, probing it with the tip of the bayonet. "Gosh, I think it's still in there!"

He planned to remove the entire burrow intact, in a block, and run it through a CT scanner back home, to see what it contained. "Any Cretaceous mammal burrow is incredibly rare," he said. "But this one is impossible—it's dug right through the KT boundary." Perhaps, he said, the mammal survived the impact and the flood, burrowed into the mud to escape the freezing darkness, then died. "It may have been born in the Cretaceous and died in the Paleocene," he said. "And to think—sixty-six million years later, a stinky monkey is digging it up, trying to figure out what happened." He added, "If it's a new species, I'll name it after you."

When I left Hell Creek, DePalma pressed me on the need for secrecy: I was to tell no one, not even close friends, about what he'd found. The history of paleontology is full of tales of bribery, backstabbing, and double-dealing. In the nineteenth century, Othniel C. Marsh and Edward Drinker Cope, the nation's two leading paleontologists, engaged in a bitter competition to collect dinosaur fossils in the American West. They raided each other's quarries, bribed each other's crews, and vilified each other in print and at scientific meetings. In 1890, the *New York Herald* began a series of sensational articles about the controversy with the headline "SCIENTISTS WAGE BITTER WARFARE." The rivalry has since become known as the Bone Wars. The days of skulduggery in paleontology have not passed; DePalma was deeply concerned that the site would be expropriated by a major museum.

DEPALMA KNEW THAT a screwup with this site would probably end his career, and that his status in the field was so uncertain that he needed to fortify the find against potential criticism. He had already experienced harsh judgment when, in 2015, he published a paper on a new species of dinosaur called a Dakotaraptor, and mistakenly inserted a fossil turtle bone in the reconstruction. Although rebuilding a skeleton from thousands of bone fragments that have commingled with those of other species is not easy, DePalma was mortified by the attacks. "I never want to go through that again," he told me. For five years, DePalma continued excavations at the site. He quietly shared his findings with a half-dozen luminaries in the field of KT studies, including Walter Alvarez, and enlisted their help. During the winter months, when not in the field, DePalma prepared and analyzed his specimens, a few at a time, in a colleague's lab at Florida Atlantic University, in Boca Raton. The lab was a windowless, wedgelike room in the geology building, lined with bubbling aquarium tanks and shelves heaped with books, scientific journals, pieces of coral, mastodon teeth, seashells, and a stack of .50-caliber machine-gun rounds, dating from the Second World War, that the lab's owner had recovered from the bottom of the Atlantic Ocean. DePalma had carved out a space for himself in a corner, just large enough for him to work on one or two jacketed fossils at a time.

When I first visited the lab, in April 2014, a block of stone three feet long by eighteen inches wide lay on a table under bright lights and a large magnifying lens. The block, DePalma said, contained a sturgeon and a paddlefish, along with dozens of smaller fossils and a single small, perfect crater with a tektite in it. The lower parts of the block consisted of debris, fragments of bone, and loose tektites that had been dislodged and caught up in the turbulence. The block told the story of the impact in microcosm. "It was a very bad day," DePalma said. "Look at these two fish." He showed me where the sturgeon's scutes—the sharp, bony plates on its back—had been forced into the body of the paddlefish. One fish was impaled on the other. The mouth of the paddlefish was agape, and jammed into its gill rakers were microtektites—sucked in by the fish as it tried to breathe. DePalma said, "This fish was likely alive for some time after being caught in the wave, long enough to gasp frenzied mouthfuls of water in a vain attempt to survive."

Gradually, DePalma was piecing together a potential picture of the disaster. By the time the site flooded, the surrounding forest was already on fire, given the abundance of charcoal, charred wood, and amber he'd found at the site. The water arrived not as a curling wave but as a powerful, roiling rise, packed with disoriented fish and plant and animal debris, which, DePalma hypothesized, were laid down as the water slowed and receded.

In the lab, DePalma showed me magnified cross sections of the sediment. Most of its layers were horizontal, but a few formed curlicues or flamelike patterns called truncated flame structures, which were caused by a combination of weight from above and mini-surges in the incoming water. DePalma found five sets of these patterns. He turned back to the block on his table and held a magnifying lens up to the tektite. Parallel streaming lines were visible on its surface—Schlieren lines, formed by two types of molten glass swirling together as the blobs arced through the atmosphere. Peering through the lens, DePalma picked away at the block with a dental probe. He soon exposed a section of pink, pearlescent shell, which had been pushed up against the sturgeon. "Ammonite," he said. Ammonites were marine mollusks that somewhat resemble the present-day nautilus, although they were more closely related to squid and octopi. As DePalma uncovered more of the shell, I watched its vibrant color fade. "Live ammonite, ripped apart by the tsunami—they don't travel well," he said. "Genus *Sphenodiscus*, I would think." The shell, which hadn't previously been documented in the Hell Creek Formation, was another marine victim tossed inland.

He stood up. "Now I'm going to show you something special," he said, opening a wooden crate and removing an object that was covered in aluminum foil. He unwrapped a sixteen-inch fossil feather, and held it in his palms like a piece of Lalique glass. "When I found the first feather, I had about twenty seconds of disbelief," he said. DePalma had studied under Larry Martin, a world authority on the Cretaceous predecessors of birds, and had been "exposed to a lot of fossil feathers. When I encountered this damn thing, I immediately understood the importance of it. And now look at this."

From the lab table, he grabbed a fossil forearm belonging to Dakotaraptor, the dinosaur species he'd discovered in Hell Creek. He pointed to a series of regular bumps on the bone. "These are probably quill knobs," he said. "This dinosaur had feathers on its forearms. Now watch." With precision calipers, he measured the diameter of the quill knobs, then the diameter of the quill of the fossil feather; both were 3.5 millimeters. "This matches," he said. "This says a feather of this size would be associated with a limb of this size."

There was more, including a piece of a partly burned tree trunk with amber stuck to it. He showed me a photo of the amber seen through a microscope. Trapped inside were two impact particles—another landmark discovery, because the amber would have preserved their chemical composition. (All other tektites found from the impact, exposed to the elements for millions of years, have chemically changed.) He'd also found scores of beautiful examples of lonsdaleite, a hexagonal form of diamond that is associated with impacts; it forms when carbon in an asteroid is compressed so violently that it crystallizes into trillions of microscopic grains, which are blasted into the air and drift down.

Finally, he showed me a photograph of a fossil jawbone; it belonged to the mammal he'd found in the burrow. "This is the jaw of Dougie," he said. The bone was big for a Cretaceous mammal—three inches long—and almost complete, with a tooth. After my visit to Hell Creek, DePalma had removed the animal's burrow intact, still encased in the block of sediment, and, with the help of some women who worked as cashiers at the Frontier Travel Center, in Bowman, hoisted it into the back of his truck. He believes that the jaw belonged to a marsupial that looked like a weasel. Using the tooth, he could conduct a stable-isotope study to find out what the animal ate—"what the menu was after the disaster," he said. The rest of the mammal remains in the burrow, to be researched later.

DePalma listed some of the other discoveries he's made at the site: several flooded ant nests, with drowned ants still inside and some chambers packed with microtektites; a possible wasp burrow; another mammal burrow, with multiple tunnels and galleries; shark teeth; the thigh bone of a large sea turtle; at least three new fish species; a gigantic ginkgo leaf and a plant that was a relative of the banana; more than a dozen new species of animals and plants; and several other burrow types.

At the bottom of the deposit, in a mixture of heavy gravel and tektites, DePalma identified the broken teeth and bones, including hatchling remains, of almost every dinosaur group known from Hell Creek, as well as pterosaur remains, which had previously been found only in layers far below the KT boundary. He found, intact, an unhatched egg containing an embryo—a fossil of immense research value. The egg and the other remains suggested that dinosaurs and major reptiles were probably not staggering into extinction on that fateful day. In one fell swoop, DePalma may have solved the three-meter problem and filled in the gap in the fossil record.

BY THE END of the 2013 field season, DePalma was convinced that the site had been created by an impact flood, but he lacked conclusive evidence that it was the KT impact. It was possible that it resulted from another giant asteroid strike that occurred at around the same time. "Extraordinary discoveries require extraordinary evidence," he said. If his tektites shared the same geochemistry as tektites from the Chicxulub asteroid, he'd have a strong case. Deposits of Chicxulub tektites are rare; the best source, discovered in 1990, is a small outcrop in Haiti, on a cliff above a road cut. In late January 2014, DePalma went there to gather tektites and sent them to an independent lab in Canada, along with tektites from his own site; the samples were analyzed at the same time, with the same equipment. The results indicated a near-perfect geochemical match.

In the first few years after DePalma's discoveries, only a handful of scientists knew about them. One was David Burnham, DePalma's thesis adviser at Kansas, who estimates that DePalma's site will keep specialists busy for at least half a century. "Robert's got so much stuff that's unheard of," Burnham told me. "Amber with tektites embedded in it—holy cow! The dinosaur feathers are crazy good, but the burrow makes your head reel." In paleontology, the term *Lagerstätte* refers to a rare type of fossil site with a large variety of specimens that are nearly perfectly preserved, a sort of fossilized ecosystem. "It will be a famous site," Burnham said. "It will be in the textbooks. It is the Lagerstätte of the KT extinction."

Jan Smit, a paleontologist at Vrije University, in Amsterdam, and a world authority on the KT impact, has been helping DePalma analyze his results, and, like Burnham and Walter Alvarez, he is a co-author of a scientific paper that DePalma is publishing about the site. (There are eight other co-authors.) "This is really a major discovery," Smit said. "It solves the question of whether dinosaurs went extinct at exactly that level or whether they declined before. And this is the first time we see direct victims." I asked if the results would be controversial. "When I saw his data with the paddlefish, sturgeon, and ammonite, I think he's right on the spot," Smit said. "I am very sure he has a pot of gold."

In September of 2016, DePalma gave a brief talk about the discovery at the annual meeting of the Geological Society of America, in Colorado. He mentioned only that he had found a deposit from a KT flood that had yielded glass droplets, shocked minerals, and fossils. He had christened the site Tanis, after the ancient city in Egypt, which was featured in the 1981 film *Raiders of the Lost Ark* as the resting place of the Ark of the Covenant. In the real Tanis, archaeologists found an inscription in three writing systems, which, like the Rosetta stone, was crucial in translating ancient Egyptian. DePalma hopes that his Tanis site will help decipher what happened on the first day after the impact.

The talk, limited though it was, caused a stir. Kirk Cochran, a professor at the School of Marine and Atmospheric Sciences at Stony Brook University, in New York, recalled that when DePalma presented his findings there were gasps of amazement in the audience. Some scientists were wary. Kirk Johnson, the director of the Smithsonian's National Museum of Natural History, told me that he knew the Hell Creek area well, having worked there since 1981. "My warning lights were flashing bright red," he told me. "I was so skeptical after the talk I was convinced it was a fabrication." Johnson, who had been mapping the KT layer in Hell Creek, said that his research indicated that Tanis was at least forty-five feet below the KT boundary and perhaps a hundred thousand years older. "If it's what it's said to be," Johnson said, "it's a fabulous discovery." But he declared himself "uneasy" until he could see DePalma's paper.

One prominent West Coast paleontologist who is an authority on the KT event told me, "I'm suspicious of the findings. They've been presented at meetings in various ways with various associated extraordinary claims. He could have stumbled on something amazing, but he has a reputation for making a lot out of a little." As an example, he brought up DePalma's paper on Dakotaraptor, which he described as "bones he basically collected, all in one area, some of which were part of a dinosaur, some of which were part of a turtle, and he put it all together as a skeleton of one animal." He also objected to what he felt was excessive secrecy surrounding the Tanis site, which has made it hard for outside scientists to evaluate DePalma's claims.

Johnson, too, finds the lack of transparency, and the dramatic aspects of DePalma's personality, unnerving. "There's an element of showmanship in his presentation style that does not add to his credibility," he said. Other paleontologists told me that they were leery of going on the record with criticisms of DePalma and his co-authors. All expressed a desire to see the final paper, which was published [on April 1, 2019] in the *Proceedings of the National Academy of Sciences*, so that they could evaluate the data for themselves.

AFTER THE G.S.A. talk, DePalma realized that his theory of what had happened at Tanis had a fundamental problem. The KT tsunami, even moving at more than a hundred miles an hour, would have taken many hours to travel the two thousand miles to the site. The rainfall of glass blobs, however, would have hit the area and stopped within about an hour after the impact. And yet the tektites fell into an active flood. The timing was all wrong.

This was not a paleontological question; it was a problem of geophysics and sedimentology. Smit was a sedimentologist, and another researcher whom DePalma shared his data with, Mark Richards, now of the University of Washington, was a geophysicist. At dinner one evening in Nagpur, India, where they were attending a conference, Smit and Richards talked about the problem, looked up a few papers, and later jotted down some rough calculations. It was immediately apparent to them that the KT tsunami would have arrived too late to capture the falling tektites; the wave would also have been too diminished by its long journey to account for the thirty-five-foot rise of water at Tanis. One of them proposed that the wave might have been created by a curious phenomenon known as a seiche. In large earthquakes, the shaking of the ground sometimes causes water in ponds, swimming pools, and bathtubs to slosh back and forth. Richards recalled that the 2011 Japanese earthquake produced bizarre, five-foot seiche waves in an absolutely calm Norwegian fjord thirty minutes after the quake, in a place unreachable by the tsunami.

Richards had previously estimated that the worldwide earthquake generated by the KT impact could have been a thousand times stronger than the biggest earthquake ever experienced in human history. Using that gauge, he calculated that potent seismic waves would have arrived at Tanis six minutes, ten minutes, and thirteen minutes after the impact. (Different types of seismic waves travel at different speeds.) The brutal shaking would have been enough to trigger a large seiche, and the first blobs of glass would have started to rain down seconds or minutes afterward. They would have continued to fall as the seiche waves rolled in and out, depositing layer upon layer of sediment and each time sealing the tektites in place. The Tanis site, in short, did not span the first day of the impact: it probably recorded the first hour or so. This fact, if true, renders the site even more fabulous than previously thought. It is almost beyond credibility that a precise geological transcript of the most important sixty minutes of Earth's history could still exist millions of years later—a sort of high-speed, high-resolution video of the event recorded in fine layers of stone. DePalma said, "It's like finding the Holy Grail clutched in the bony fingers of Jimmy Hoffa, sitting on top of the Lost Ark." If Tanis had been closer to or farther from the impact point, this beautiful coincidence of timing could not have happened. "There's nothing in the world that's ever been seen like this," Richards told me.

ONE DAY SIXTY-SIX million years ago, life on Earth almost came to a shattering end. The world that emerged after the impact was a much simpler place. When sunlight finally broke through the haze, it illuminated a hellish landscape. The oceans were empty. The land was covered with drifting ash. The forests were charred stumps. The cold gave way to extreme heat as a greenhouse effect kicked in. Life mostly consisted of mats of algae and growths of fungus: for years after the impact, the Earth was covered with little other than ferns. Furtive, ratlike mammals lived in the gloomy understory.

But eventually life emerged and blossomed again, in new forms. The KT event continues to attract the interest of scientists in no small part because the ashen print it left on the planet is an existential reminder. "We wouldn't be here talking on the phone if that meteorite hadn't fallen," Smit told me, with a laugh. DePalma agreed. For the first hundred million years of their existence, before the asteroid struck, mammals scurried about the feet of the dinosaurs, amounting to little. "But when the dinosaurs were gone it freed them," DePalma said. In the next epoch, mammals underwent an explosion of adaptive radiation, evolving into a dazzling variety of forms, from tiny bats to gigantic titanotheres, from horses to whales, from fearsome creodonts to large-brained primates with hands that could grasp and minds that could see through time. "We can trace our origins back to that event," DePalma said. "To actually be there at this site, to see it, to be connected to that day, is a special thing. This is the last day of the Cretaceous. When you go one layer up—the very next day—that's the Paleocene, that's the age of mammals, that's our age."

UPDATE

Tanis continues to yield extraordinary and astonishing fossils, including an intact pterosaur egg with the embryonic skeleton still inside and a piece of a Thescelosaurus *dinosaur with the skin beautifully fossilized and with enough remaining organic material that they might be able to determine the actual coloration of the skin. DePalma found two pristine fragments of the original asteroid embedded and preserved in amber. The site will continue to yield invaluable information for a century or more. Tanis and DePalma were the subject of an excellent BBC documentary,* Dinosaurs: The Final Day, *narrated by David Attenborough, which aired in 2022.*

It still sends shivers down my spine to think that our very existence is due to a random piece of space junk—a god rock, if you will—which had been wandering through the solar system for billions of years, colliding with our planet. There's a profundity in that idea beyond words.

CURIOUS CRIMES

THE CLOVIS POINT CON

THE GREATEST HUMAN hunters to walk the planet may have been the Clovis, a mysterious people who appeared abruptly in North America about thirteen thousand years ago. The Clovis were among the first of our species on this continent to develop weaponry that allowed them to go one-on-one against big, dangerous game. Their favored prey was the Columbian mammoth, the largest land animal that Homo sapiens has ever hunted in the New World. It stood as high as thirteen feet at the shoulder, weighed between eight and nine tons, and could run perhaps twenty-five or thirty miles an hour. It was fifty percent heavier than the African elephant, which is today a dangerous animal to hunt even with a high-caliber rifle. Like the elephant, a provoked mammoth probably slashed at and gored its assailant with its tusks; trampled, crushed, and battered him with its head; and, picking up what was left, flailed him about until the pieces flew apart in bloody gobbets.

The Clovis hunters stalked mammoths with two particularly lethal weapons. The first, a thrusting spear, was for close-in work. The second may have been an atlatl—a spear launched with a throwing stick. (The bow and arrow weren't invented until thousands of years later.) Both types of Clovis spears were tipped with a long, tapered projectile point flaked from stone. The Clovis point was beautifully designed to kill a mammoth. Viciously sharp, and unlikely to break against the

---

Originally published in the *New Yorker* in 1999 as "Woody's Dream."

hide, it tore a big hole in the tissue and caused heavy blood loss. Every time the mammoth took a step after being hit, the point would saw and slice the flesh.

How the Clovis hunted mammoths is disputed. They may have waited until a young animal strayed from its family group, hit it with a few atlatls, and followed it until it died. They also may have killed a full-grown mammoth by getting up beside it and shoving spear after spear through the ribcage into the vital organs. Either way, a mammoth took a lot of killing: one skeleton, found in southern Arizona, had eight Clovis points in its body—and that mammoth may have been one that got away.

The Clovis traveled hundreds of miles to find gorgeous, semiprecious stones suitable for their points: translucent agate, chert, smoky-quartz crystal, red jasper, obsidian. These stones were not just beautiful to look at; when knapped (chipped), they formed keen edges and glossy flake scars that were almost silky to the touch. To make pleasing flake patterns on the stones' points, the weapons producers used a technique that the French call *outré passé*, or overshot flaking. To carry out this exceedingly difficult procedure, the edge of the stone was struck in such a way that a long flake, shaped like a scimitar, traveled all the way across the face of the spear point. Sometimes the Clovis knapped the stone on the diagonal, for further decorative effect. When you hold a fine Clovis projectile point—maybe five inches long, leaflike, cool, deadly, balanced, luminous—you know that you are holding something more than a tool. It reaches beyond the utilitarian into the realm of aesthetics—each, like a piece of Shaker furniture, is an object made with skill, love, simplicity, and faith.

About nine or ten thousand years ago, the Clovis and their Paleo-American descendants disappeared—no one knows how or why—and the high art of stone-knapping disappeared with them. Nothing like it was achieved again for thousands of years. Even today, flint-knappers cannot re-create a Clovis point perfectly. Clovis points are rare, sexy, and extremely valuable.

ONE OF THE world's foremost collectors of these artifacts is a trim Texan named Forrest Fenn. Now in his sixties, Fenn, a former Air Force pilot who was shot down over Laos during the Vietnam War, lives in Santa Fe, in a large house hidden behind adobe walls. Wagon ruts from the Santa Fe Trail run across his back yard. In the early seventies, Fenn started the first major art gallery in Santa Fe, and he was the dealer who established Santa Fe as an international art market. Fenn sold the gallery in 1988, and has since spent most of his time publishing books and excavating an ancient five-thousand-room Indian ruin he owns outside the city. He has never outgrown a boyish passion for Indian artifacts, fossils, bones, and arrowheads. His hoard of Clovis points is one of the most outstanding private collections in the nation.

One day last fall [1998], Fenn, who is a neighbor and an old friend of mine, called me up. Nine months earlier, he said, he had bought three Clovis points from an antique dealer based in Rollins, Montana, for fifteen thousand dollars. Then, a few weeks ago, the dealer had arrived on his doorstep with nine more, and wanted to sell them for a hundred and thirty-five thousand dollars. Eventually, the dealer dropped the price to eighty-five thousand—an excellent figure—and Fenn took the points on approval. He invited me over to see them.

Fenn's various collections are displayed around his office—a fossil skull of a giant short-faced cave bear, a mummified falcon from Egypt, old Indian pots, beaded moccasins, arrowheads, peace pipes, tomahawks. I found him at his kitchen table, with the twelve spear points spread out on black velvet. They ranged from three inches long to almost seven. Most were made of chert or jasper; two had been flaked from pure quartz crystal. The points, the dealer told him, had been found near Greeley, Colorado, in the 1920s, by a ranch foreman named Tom Jenson; most of them had turned up recently, after Jenson's death, stashed inside his house.

"When I first saw those points," Fenn said, "I felt like Dr. What's-His-Face when he looked through that hole and saw King Tut lying in his sarcophagus." In particular, Fenn felt that the longer quartz-crystal point was surely one of the greatest Clovis artifacts ever found. The points bore a strong resemblance to the celebrated Drake points, a cache of thirteen spear points found in 1978 in a wheat field in the same part of Colorado. (Most of those are now in the Smithsonian.) Fenn wondered if these points might have come from the same cache—and perhaps have even been flaked by the same Clovis artist. If so, that would make them still more important. The points also appeared to bear minute traces of red ochre, which is a kind of iron oxide; the Clovis, for unknown reasons, often packed caches of their points in this material.

Fenn told me that two of the world's foremost authorities on the Clovis—George Frison and Robson Bonnichsen—had just come to see the points. Frison, a professor emeritus at the University of Wyoming, which has named its George C. Frison Institute of Archaeology and Anthropology after him, is a member of the National Academy of Sciences, and is the top authority on Clovis hunting. Bonnichsen is the director of the Center for the Study of the First Americans, at Oregon State University. Both men loved the points. "When I first saw them, I thought they were wonderful," Frison told me later. "I couldn't believe that crystal point: it looked perfect." Bonnichsen also believed that the points were genuine. "Not only was the form of the artifacts right," he said, "but the technology was right"—the points had been made using Clovis flaking and reduction techniques.

Fenn had sent the first three points to an authenticator—Calvin Howard, in Springfield, Illinois—and Howard examined the points using visual, optical, microscopic, and ultraviolet tests. Howard said that although there was no foolproof way to verify the points' authenticity he was nevertheless willing to declare them authentic—to "paper" them, in the jargon of the trade. In this field, authentication often hinges on subjective judgment. These points *looked* right. No modern flint-knapper, it was generally believed, had ever been able to replicate the Clovis touch. Even so, the sheer brilliance of the points gave Fenn pause, for only once in a generation did Clovis points as good as these appear. For a hundred thousand dollars, he had to be sure. "I climbed in my airplane," he told me later, "and took these things to Idabel, Oklahoma, through terrible rainstorms." He wanted to show them to Gregory Perino, an eighty-five-year-old man who, he said, "probably knows more about Indian artifacts than anybody else alive" and could definitively authenticate them. When Perino saw the Clovis points, he, too, was impressed, though he noted some red clayey material on the points which made him slightly uneasy. The one thing he didn't like was a microscopic film he detected on their surface, which looked like Teflon. But this didn't cause him undue concern, since many collectors oil or wax their points. The presence of the film troubled Fenn, however. He sent one of the points—and, later, all of them—to Kenneth Tankersley, an assistant professor of anthropology at Kent State University, in Ohio, who is an expert on microscopic and chemical analysis of Paleo points.

"They were absolutely breathtaking—spectacular," Tankersley told me. "There was no question in my mind that this was the Clovis technology. My words to Forrest were 'This is as good as gold.'" Fenn asked Tankersley to run a battery of tests anyway. At the same time, other extraordinary Paleo projectile points began to appear on the market. Made by descendants of the Clovis, who hunted a large buffalo, *Bison antiquus* (now extinct), the points were impeccably designed for penetrating the skin of that shaggy, muscular animal. Usually thinner, shorter, and smoother, these later point styles have been given such names as Folsom, Alberta, Scottsbluff Firstview, Dalton, and Eden, after the places where they were originally found. One of the country's top dealers in Paleo points—Jeb Taylor, of Buffalo, Wyoming—bought three that were on the market: an Alberta and two Folsoms. "The Alberta point was absolutely right," he told me recently. "It really struck me as being unfakable." (Taylor had had the points authenticated, just to be sure.) He paid six thousand for the large Folsom, three thousand for the Alberta, and fifteen hundred for the other Folsom—all excellent prices. Taylor sold the lesser Folsom for thirty-five hundred, the Alberta for sixty-five hundred, and kept the larger, and better, Folsom point for himself.

Toward the end of 1998, Forrest Fenn got a call from Kenneth Tankersley, who was in the middle of his battery of tests. He had uncovered a number of problems. For one thing, the red ochre on some of the points was too bright: usually, an iron-eating microbe degrades the ochre over time, mellowing the color. The other points were stained not with ochre but, suspiciously, with red dirt. (The particular points that Calvin Howard had examined were not stained.) Looking through a microscope, Tankersley could see tiny flakes still adhering to the stone—flakes that should have popped off after thirteen thousand years of freezing and thawing. Even more problematic, he told Fenn, were the two quartz-crystal points: the quartz had rutilation (embedded crystal hairs), which strongly suggested that the stone came from Brazil. He asked Fenn for permission to perform more extensive chemical tests. Fenn granted it.

Tankersley examined the film that was coating the points. He rubbed a flake scar on each point with a Q-Tip dipped in solvent, teased a bit of the coating off, and viewed it under ultraviolet light. It appeared to be some kind of modern petrochemical derivative—dirty motor oil, he guessed. "I thought, What if the discoverer was out in a tractor and dumped them in the back, where there was oil? I wanted to make absolutely sure. Two of the top people in the country, Robson Bonnichsen and George Frison, had seen the points, and the points had looked good to them. These are two colleagues very senior to me. For me to say, 'Wait a minute, these are fakes!' I wanted to have the kind of evidence that you would need in a court of law for a murder case—beyond a reasonable doubt." One of the points was made of a material that Tankersley recognized as Knife River flint. This flint has a distinctive property: when it is freshly flaked, it usually fluoresces yellow under long-wavelength ultraviolet light, but if the flaking is very old it tends to fluoresce orange or green. Tankersley cleaned one flake scar and hit it with UV light. The scar fluoresced yellow.

"One of the hardest things I ever had to do," Tankersley told me, "was to call Forrest on the telephone and tell him, 'I think these are fakes.'"

Fenn remembers the conversation vividly. "I felt like the director of the Louvre when scientists tell him that the Mona Lisa is a fake," he said. "These were the Mona Lisa of points."

Fenn, of course, immediately began to wonder who might have faked them. He didn't have to think long.

WOODY BLACKWELL LIVES in a small bungalow in Warner Robins, Georgia. He drives a cheap Toyota truck and has what he calls a "frugal" lifestyle. He was once a counterintelligence specialist in the Air Force, and he still looks it: a fit man of forty-two, he is neat in appearance, stocky, and wears a close-cropped beard in the winter months. He carries himself with a military bearing, and his words are straightforward and precise.

When Blackwell was seven, he was digging with some friends one day in his back yard, in Millbrook, Alabama, and discovered an arrowhead—his first. "It was as if a switch was flipped," he told me recently. Ever since, he said, he has been obsessed with flint-knapping. "Our species and others like it have been banging on rocks for two and a half million years. It's only in the last couple thousand that we've been doing something else. When you rediscover it—that feeling of banging on rocks—you *have* to do it. It's almost a need. Maybe it's genetic."

His first attempts at knapping were a disaster. "My tool kit consisted of a sixteen-penny nail and a piece of a brick," he recalled. "I just gouged up my hands and lost a couple of fingernails and made a mess of things."

Nine years ago, when Blackwell was still in the Air Force, he fell in with some Texas flint-knappers and took up the hobby seriously. He joined what is in America a perfervid subculture. In the 1970s, knapping was a rarefied hobby: its few aficionados would get together on Sundays to trade techniques about the lost art. By the time Blackwell got involved, it had become something of a craze. Now there are more than a hundred "knap-ins" around the country every year, and some are attended by thousands of people. At one recent knap-in, in Missouri, participants consumed twenty truckloads of rock as they created countless arrowheads. There are dozens of websites devoted to the hobby. D. C. Waldorf, who was an early enthusiast, has sold more than forty thousand copies of a book he wrote in 1975 entitled *The Art of Flint Knapping*. There are thousands of knappers now working around the country, turning out an estimated million and a half stone points and tools every year. An article in the journal *American Antiquity* said that this country is experiencing "a twentieth-century stone age."

Knappers, it must be said, command almost no respect in the archaeological world. Most experts, if they think of flint-knappers at all, think of redneck hobbyists pounding rocks in the backwoods or of people who stock cheesy Indian souvenir shops at gas stations along the highways of the West. Knapping is hardly considered a craft, let alone a fine art.

Nevertheless, over the past four years Blackwell eked out an income as a professional flint-knapper, selling replica points for fifty or a hundred dollars apiece. It's a poor way to make a living: in his best year, 1996, Blackwell grossed less than twenty-two thousand dollars. Sleazy dealers often bought points from him. He sometimes saw his own replicas doctored up to look old and displayed at arrowhead shows as authentic artifacts worth hundreds, or even thousands, of dollars more than he had sold them for.

From the beginning, Blackwell revered Clovis points. He collected originals, and when he couldn't afford to buy real points he acquired museum-quality casts. He traveled around the West, collecting the same stone that the Clovis used. He looked at the Drake Cache, at the Smithsonian, and at other Clovis caches, and, through trial-and-error experimentation, deduced complex Clovis flaking and reduction techniques. He grew intimate with all facets of Paleo technology: its polyhedral cores, bifacial flaking, preforms, spalls, and isolated platforms. And then he began to reverse-engineer the Clovis point.

"There is a feel about old Clovis points," he told me recently. "They aren't fussed over. They are everything they need to be, and nothing more." Blackwell also began reverse-engineering other styles of Paleo points: Folsom, Alberta, Scottsbluff, Firstview, and Eden. If anything, these points are even more difficult to copy than Clovis points are. (Some of the minor details of pressure-flaking on an Eden or a Folsom point, for example, can be seen only with a microscope.) "They are sublimely functional," Blackwell said. "And yet the beauty, the workmanship, the artistry is so far past the function that it's almost religious. They're no longer just a sharp edge to stuff inside a buffalo. Those guys made the most incredible pieces you can imagine. And I think I've figured out how."

It takes Blackwell two or three hours to make a Clovis point, and up to eight hours to make an Eden point. Many break, and many others are aesthetic failures. Moreover, knapping points is a dangerous business: almost every point will draw its maker's blood. About once a month, Blackwell cuts himself deeply enough to need stitches, but instead of getting them he stops the bleeding with spiderwebs used as a styptic (a folk remedy from his Southern boyhood), or just superglues the wound shut.

In 1994, Blackwell began sending his replicas to museums and archaeologists, and offering to share with them what he had rediscovered about Paleo-American stone technology. Most never bothered to respond. One museum asked him to make two hundred Clovis replicas for twenty dollars each, a price that would hardly pay for the cost of his stone. It became clear to him that he had dedicated his life to an unremunerative muse.

A few years ago—Blackwell says he can't remember exactly when—he was working on a Clovis point, using a favorite Clovis material: Alibates chert, from Texas. After a few hours, he paused to examine his work. He says that he had nothing sinister in mind. "I just stopped and looked at this piece and said, 'That really looks like a Drake-style Clovis if I stop right here.' Until then, I had always kept going, cleaning up the edges, making the point smoother, getting the symmetry dead on, and really dressing the thing up. What I'd been losing was its immediacy, its simplicity." The Drake point is lovely and distinctive: long, narrow, almost parallel-sided, with multiple "fluting"—long vertical flakes struck from the base.

During the following month, Blackwell made two more Drake points, again striving for simplicity and avoiding carrying the work too far. For inspiration, he reflected on his ancient predecessors. "I tried to get in their heads. What were their ideals? And I tried to re-create that." He noticed that these points, too, were very, very good.

Drawing on an encyclopedic knowledge of Clovis points, he

added small, verisimilitudinous details. He ground off the sharp edges around the base of each point—something that the Clovis had done to keep the stone from cutting through the sinew used in hafting and to prevent breaking. He sawed through heavy leather and through antler bone with the points, to simulate the edge wear caused by the killing and butchering of mammoths. He then resharpened some of the dulled points, just as the Clovis would have done.

Blackwell had achieved a certain kind of perfection; these were small masterpieces. Museum curators might scoff at him, but, as Blackwell said to me later, the finest Clovis artist "would have been entirely happy to strap one of these on his foreshaft and go kill a mammoth."

It was the next step that proved Blackwell's undoing. For all the points' brilliant workmanship, Blackwell realized, they still looked new. As an experiment—still without nefarious motives, he says—he decided to age them artificially. He put them, along with some dirt from his back yard, in a tumbler—a motorized drum that turns stones for polishing. He let the tumbler rotate with the dirt for two or three days. (Unbeknownst to Blackwell, it appears that the plastic lining of the drum shed hydrocarbons into the mixture, giving the points a Teflon-like coat.) When the points came out, some of them still looked a little bright. Blackwell rubbed a small amount of potassium permanganate on them, to darken them, and he smeared red ochre on some points—ochre that he had collected from the Bear Mouth ochre deposit, in Montana, on one of his trips. When that ran out, he substituted red Georgia dirt similar in appearance but chemically different.

The "old" points sat around in his shop for weeks while his financial troubles mounted. "I was desperate," he says.

In late 1997, by his account, he first began mentioning the points in passing to collectors and dealers. "I really don't know when or why or how I decided to do it," he says. "I just started talking to guys. I don't know if I was trying to work up the nerve to do something with them or trying to talk myself out of it. I just asked around, saying, 'I've got three Clovis points, two good ones, and one little beat-up guy. How much could I get?' And they said, 'Oh! Where *are* they? Let's see the pictures!' And it took on a life of its own. I don't want to give myself an out here. But there were guys who were all but reaching through the phone line and grabbing me to pull me into their living rooms. These guys are rabid. It's like they need their next fix."

In late March 1998, Blackwell ran into Forrest Fenn at an arrowhead show in Collinsville, Illinois, and mentioned that he knew a man in Montana, an antique dealer, who supposedly had some Clovis points for sale. Knowing that Fenn might not buy points from a knapper, Blackwell had arranged for this friend to front for him. (Everyone agrees that the antique dealer was duped by Blackwell; the man has a severe heart condition, and I was asked not to mention his name in the article for fear of giving him a heart attack.) Fenn snapped them up, paying the antique dealer fifteen thousand dollars and giving Blackwell an additional fifteen hundred dollars as a "finder's fee." This was such easy money that Blackwell quickly began flaking nine more points to sell to Fenn.

Blackwell says that he wanted his points to be the "heartbeat" of Fenn's collection—a tall order when you consider that Fenn owns some of the greatest Clovis points ever made. It was easy for Blackwell to study Fenn's collection surreptitiously: Fenn makes his material available for scientific scrutiny; his collections have been widely published; and casts of many of his points are in general circulation. The centerpiece of Fenn's collection is a cache of Clovis points—the Fenn Cache—from the area where Utah, Idaho, and Wyoming meet. In this cache is a singular point made of an extremely rare and beautiful material called phosphoria, which is found in the Bighorn Mountains, in Wyoming. This spear point displays possibly the finest diagonal *outré passé* flaking in existence. Blackwell acquired a cast of this point and studied it intensively. One of the points that Blackwell sold to Fenn was made of phosphoria, which he had found after an exhaustive search in a riverbed below the Bighorns. He had flaked it using the same *outré passé* diagonal technique. The Fenn Cache contains three quartz-crystal points, and Blackwell also duplicated these with his crystal fakes. By giving the dozen points a Drake "look," it appears, he may have been tempting Fenn further: only thirteen points had come from the Drake Cache, and many believe that there are more to discover.

Blackwell courted other buyers. According to a major collector, Mark Mullins, Blackwell often cornered collectors and dealers at arrowhead shows around the country and casually questioned them about their taste in points. What stone did they prefer? What localities and types did they like best? He delivered to them what they most wanted, and then further romanced his victims with elaborate stories about how and where the points had been found.

"Stories sell points," Blackwell told me. "The more outlandish the story, the more desirable the point." Blackwell made up a story about the Alberta points he had sold to Mark Mullins. It involved a Canadian paratrooper, named Oscar, who had just returned home from the Second World War. To back up the story, Blackwell supplied a letter that Oscar had supposedly written when he sold Blackwell the points. "I came home from W.W. II in April 1946," the letter stated. "Went hunting that fall in the Sandhill Country of Saskatchewan." As he was wandering in the backcountry, Oscar wrote, he came across a litter of old chalky bones mingled with "arrowheads." He picked up eight whole ones. "When I got home, I wrapped the arrowheads in a *Life* magazine. Still remember how mad my mother got cause she hadn't read it yet." The points that Mullins received were duly wrapped up in pages torn from a 1946 issue of *Life*. Blackwell told me he hadn't been "gunning" for Fenn, Mullins, or anyone else. "I wanted them to feel like they were getting their money's worth," he said.

Nevertheless, he told me, he soon began to feel deeply troubled about forging the points. "What I wanted very badly was to provide a better life for my family. I did a desperate thing in a desperate situation. I didn't like it when I did it, and I like it even less now." He paused, and then added, "There are nights when I wish I'd never found that first point when I was seven years old."

Later, Blackwell described his agony to me: "I'm a law-and-order kind of guy," he said. As a military officer, he had always placed a high value on being what he called "an honorable man." Now he could no longer call himself that, and, he said, "It makes me feel small." At one point, while he was watching the Presidential-impeachment hearings, he said, a terrible thought struck him: "I'm no better than Bill Clinton!"

AS SOON AS Forrest Fenn heard Kenneth Tankersley's unpleasant news about the authenticity of his new points, he recalled that it was Woody Blackwell who had told him about the antique dealer in Montana with the Clovis points for sale. When he discovered that his fellow collector and friend Mark Mullins had bought some Alberta points from Blackwell that had the same Teflon-like coating on them, it confirmed his suspicions. Jeb Taylor had also bought his points from Blackwell.

"I called Blackwell up," Fenn said. "I told him, 'These points are traveling a lot of roads and they all lead back to you.'"

Blackwell, for his part, remembers it differently: after a concerned call from Mark Mullins, he called Fenn on his own initiative, he says, to confess, and told him, "Mr. Fenn, the points you've got are not authentic, and I know that because I made the things." Blackwell says that he was unaware of the furor going on about his fake points and did not know that he was on the verge of being exposed. Taylor has his doubts. "I said to Woody, 'Well, if it was conscience that made you confess, it was pretty good timing, because you were a week away from being nailed.'"

Whether out of conscience or in a panic about getting caught, Blackwell acted quickly. He picked up the telephone, and over the course of a humiliating week he called everyone who had been affected by what he had done. He called the authenticators and described exactly how he had made the points. He called collectors and friends to apologize. He wrote frank letters of acknowledgment to all those he had dealt with. He FedExed everyone's money back—a move that was very painful for him financially.

Some collectors didn't want to believe him. When he told Jeb Taylor that he himself had made the points, Taylor started to protest. "My reaction was 'Not the Alberta!'" Taylor recalls. "I was about to contest him, really. I just felt so sure about that point." Taylor stopped when he realized how ridiculous it was for him to argue against the faker himself.

Despite the interstate nature of Blackwell's questionable financial transactions, nobody reported him to the FBI. Since everyone had got his money back, no one, apparently, felt inclined to pursue a prosecution.

"I wasn't hurt," said Fenn, "and I felt all along that his motivation wasn't money."

"I was really pissed about it," said Taylor, "but at the same time I respected his ability. I actually congratulated him."

How was Blackwell able to achieve something that so many experts had said was impossible? And why had he sold the points directly to the country's premier collectors and experts, instead of seeking out rubes and suckers in the hinterlands? "He could have made penny-ante thousand-dollar points," said Mullins, "and sold them to many people who wouldn't have gone to the trouble to get them authenticated, circulated, and examined."

The other day, I dropped by Fenn's house to take another look at the fake points. He brought them out, half a dozen in each fist, clinked them down on the bare kitchen table, and spread them carelessly around: no more caresses and black velvet. He picked out the remarkable quartz-crystal point and held it up to the New Mexico sunlight streaming through the window. He waggled it, strafing the room with flecks of light. "Everyone said it couldn't be done," he said, with a grin. "I think his motivation was to show these experts that he *could* do it. And he did."

WHEN WOODY BLACKWELL first began trying to reverse-engineer the Clovis point, he had a dream. It was, he says, the most remarkable dream of his life. It was night, and he was walking through some brush. Up ahead, he could see a group of men sitting in a circle around a campfire. They were wearing cloaks or skins. They were passing things around the fire, from hand to hand, admiring them. When he got close, he could see that the things were big, beautiful Clovis points. A man held one up to the firelight. Blackwell recognized it as a huge point from a cache discovered in East Wenatchee, Washington. The light from the fire was shining through it, setting it aglow. He tried to take a step closer to see better, and made a noise. The man turned and smiled at him. "I'm not a mystic," Blackwell says. "I'm not a religious man, and I'm not one of those pudding-headed New Agers out of Sedona. But, all the same, every Clovis point I've ever made, I wonder if *this* will be the price of admission, the ticket to let me go back and sit with those guys."

UPDATE

Forrest Fenn eventually sold his arrowhead collection, including the Fenn Cache, to a private collector for more than a million dollars, possibly the largest sum ever paid for Native American arrowheads. In 2010, as most of the world knows, Forrest hid a treasure chest of gold somewhere in the Rocky Mountains and issued a poem, called "The Thrill of the Chase," which contained clues to the treasure's location. The treasure was found in Wyoming in 2020 by Jack Stuef, a medical student. Forrest passed away in September 2010. Much of the Fenn treasure was sold off piecemeal at auction in December 2022, netting $1.3 million. Woody Blackwell continues to knap extraordinary points, and his reproductions are highly sought after by collectors.

TRIAL BY FURY

The bitch needs to die naked tasting her own blood.

Your daughter will come out of prison a hard nosed lesbian (with her sex drive) I hope you are dead before that happens I really do. Meredith Kercher is dead and you dont give a shit about her or her family. But you are the victims arent you? Rot in hell.

And Raffaele? He's such the pervert he needs to be locked away for life. . . . Raffaele was perfect for Amanda infusing the sort of wicked and wonderful temptation of evilness that comes with pranks and mutual masturbation. They know nothing of love and compassion but only of spontaneous and twisted self gratification.

There are a whole lot of women who instinctively think she is a total fake, has not fooled any one of us, believes she is foul to the bone and we hope she rots in prison and dies in hell.

I hope Knox stays in jail where she is safe from me and others like me because if she ever makes it home to Seattle she will suffer that same fate as her victim. None of you can stop me either.

Originally published as a Kindle single in 2013.

On November 2, 2007, in the ancient and lovely hill town of Perugia, Italy, a British girl named Meredith Kercher was found murdered in the cottage she shared with several other students. Her half-naked body lay on the floor of the bedroom, covered by a duvet; her throat had been stabbed, and there were signs of sexual assault and robbery. It was one of the most brutal murders in Perugia in more than thirty years, and it made front-page news in Italy. Four days later, police and prosecutors held a triumphant press conference in Perugia, in which they proclaimed that they had solved the case and arrested the killers. "Case closed," they announced. Among the three alleged killers were a twenty-year-old college student from Seattle named Amanda Knox and her Italian boyfriend, Raffaele Sollecito. The ensuing investigation, trial, conviction, and appeal of Amanda and Raffaele lasted 1,428 days and became one of the most sensational murder trials of the new century. On October 3, 2011, an appeals court in Perugia declared them innocent. Amanda went back to her family in Seattle and Raffaele to his in Bari. On March 26, 2013, Italy's Court of Cassation vacated the acquittal and ordered a retrial on points of law as yet unspecified. The retrial may well involve more years of appeals and reviews. Amanda was arrested when she was twenty; she could be thirty by the time her case is finally resolved.

Within days of Amanda's arrest, public opinion began lining up on either side. It eventually coalesced into two groups, the so-called "Guilters" and the "Innocentisti," anti-Amanda and pro-Amanda bloggers.

These two groups have been brawling online ever since. People sometimes note the transient, ever-changing nature of the Web, but in fact the opposite is true. The Web is a gigantic tar pit that traps and fossilizes every electron that ventures within. On March 29, 2013, as I was putting the final touches on this article, I conducted an experiment. I Googled "Amanda Knox" and got 7.1 million hits. I then tried "Amanda Knox" and "bitch," which returned 1.7 million hits. "Amanda Knox" and "pervert" came back at 880,000 hits, and her name coupled with "slut" yielded 380,000. "Amanda Knox" and "innocent" returned 482,000. The quotations that opened this article were gathered in about fifteen minutes of surfing. There are millions of similar comments about Amanda like this, and most of them will survive in digital form for a long time—perhaps, given Web archiving efforts, close to forever. Amanda's great-great-grandchildren may find that this ugly archive is only a few clicks away.

The extreme viciousness of the anti-Amanda commentariage is startling. There are countless statements calling for the murdering, raping, torturing, throat-cutting, frying, hanging, electrocution, burning, and rotting in hell of Amanda, along with her sisters, family, friends, and supporters. This silicon Inquisition is still there in all its glory, undiminished by time.

Which brings me to the question: Why did the Knox case arouse such a furor on the Web? And this leads to an even more interesting problem: Why are there so many savage, crazy, vicious, and angry people on the Internet? The answer, which might appear obvious on the surface, is in fact anything but clear. Recent controversial research into the evolution of altruism, warfare, and punishment in human society indicates that Internet savagery may be programmed into our very genes.

I WAS DRAWN into the case by accident. Amanda's chief prosecutor was Giuliano Mignini, a man I knew well. In 2000 I moved to Italy with my family to write a murder mystery set in Florence. We settled in a fifteenth-century farmhouse in the Florentine hills. I soon learned that the picturesque olive grove outside our door had been the scene of a horrific double homicide in 1983, committed by a serial killer known only as the Monster of Florence. Between 1974 and 1985, the Monster killed young people making love in cars in the Florentine hills and performed a ritualistic mutilation on the women's bodies. He had never been identified, and the case, one of the longest and most expensive criminal investigations in Italian history, had never been solved. The Monster was so depraved, and so skilled at murder, that he made Jack the Ripper look like Mister Rogers. I became interested and dropped the idea of the novel to write a book about the Monster case instead. I teamed up with the Italian journalist Mario Spezi, who had covered the Monster's killings for the local paper from the beginning and knew more about the case than even the police.

Giuliano Mignini did not like our investigation. It went against his theories that a satanic cult was responsible for the Monster killings—despite clear forensic evidence that all fourteen victims had been killed by a lone individual. He launched a secret investigation of us, tapped our cell phones, and bugged Spezi's car. He had the police seize Spezi's computer and all our notes, research, and files on the case. The police then picked me up on the streets of Florence and hauled me in before Mignini, where he interrogated me for hours, with no attorney or interpreter present. He demanded I confess to a string of crimes, including being an accessory to murder, and when I refused, he indicted me for perjury and obstruction of justice and suggested I leave the country. Spezi fared worse, much worse. Mignini ordered him arrested and accused him of *being* a member of the satanic sect that conducted the Monster killings. Spezi was thrown into the same prison in which Amanda Knox would later be incarcerated. Spezi remained jailed until an international uproar, led by the Committee to Protect Journalists, forced his release. Together Mario and I published a bestselling book about the case, *The Monster of Florence*, which is now being made into a television series. (The Italian courts dismissed all charges against us; Mignini was indicted and convicted for abuse of office, the conviction later suspended on a technicality.)

A few days after Amanda Knox was arrested for murder, I got a call from a man named Tom Wright, a former Hollywood executive and well-known filmmaker. His powerful voice, full of desperation and breaking at times, came booming down the wire. He explained that his daughter and Amanda were high school friends and schoolmates. He knew her family. It was impossible that Amanda could be a murderer. He had heard about our book and, seeing that Spezi and I had also been victims of Mignini, begged us for help.

I wasn't sure about Amanda's innocence at the time, but when I looked into the case, I was shocked. Mignini and the Perugian police were railroading Amanda and Raffaele for a murder they did not commit. Spezi and I later learned why she had been framed, which we detailed in a new afterword to *The Monster of Florence*, published on April 23, 2013.

I felt like I had to become involved.

My first foray into a public discussion of the case was in an interview with the journalist Candace Dempsey, who wrote a blog hosted by seattlepi.com, the website of the *Seattle Post-Intelligencer* newspaper. Dempsey had been the first journalist to raise questions about the case against Knox. She warned me ahead of time that I would be attacked by anti-Amanda bloggers. I confidently assured her that, as a novelist and journalist for thirty years, I was fully hardened against bad reviews and negative comments. She posted the interview on February 8, 2008, in which I told of my own experience with Mignini and said I thought Amanda and Raffaele were innocent. It was a mild interview in what I assumed to be a rather obscure corner of the Web.

Then the comments poured in. I was stunned at their ferociousness against Amanda. But what surprised me even more were the blazing personal attacks against me. The commentators had researched me on the Web and extracted personal details I had no idea were there. They threw back at me my own biography, twisted beyond recognition, along with quotes from bad reviews of my books and ugly references to my family. They claimed that my interest in Amanda was sexual. They said I was mentally ill. They said I was a racist. Dempsey deleted the offensive postings and locked her blog at night, which only aroused the bloggers more and sent them seeking other sites to vent their fury.

Like a damned fool I waded into the fray, posting in the comments section, defending myself, attacking my attackers, and countering their criticisms. I had my name on a Google alert, and the alerts began pouring in, directing me to attacks appearing elsewhere. I found myself swept up in the drama, obsessively checking the Web multiple times a day, outraged and panicked that the accusations, especially the sexual ones, would remain on the Web forever, read by my children and unborn grandchildren. I had to answer each one, get my licks in, set the record straight; but the more I fought, the more the tide of vituperation came back at me. I felt like King Canute trying to turn back the sea with his sword. For days I was in a frenzy.

Finally, I came to my senses. I couldn't believe that I had gotten sucked in and become almost as crazy as they were. But it made me wonder: Who *are* these people? And why would so many people, unconnected with either the victim or the accused, with no skin in the game, devote their time and energy to seeing this girl punished—and to vilifying all those who came to her defense? One could understand the single-minded fervor of Amanda's family and friends in defending her. And one could appreciate the passion of those who thought she was innocent and sought to correct a terrible injustice. But why the white-hot zeal from apparently random people to see her *punished*? There was no equivalence between Amanda's defenders and her persecutors. The former were engaged in normal human behavior, the latter in something that felt pathological.

When you ask Web sophisticates why people are so vicious on the Internet, you get a set of stock responses. The very question is naïve. What do you expect? The world is full of angry people who don't have a life. The Web offers a perfect outlet where they can be anonymous, important, and powerful, and attack others without fear of retribution. The Web has given them a voice when before they had none. These are people who find meaning in their lives by connecting with similar people on the net, who seek a sense of purpose and fulfillment online that they can't achieve in the real world. Finally, the nature of the Internet, we are told, is also to blame—it's a place where the human id runs amok, it's a playground for disturbed people, it's an echo chamber for the uninformed. We are advised that Internet nastiness is white noise, best ignored. It has little effect in the real world.

While many of these explanations are undoubtedly true, none go deep enough. None explain why the Web is a place where some human beings devote enormous effort to attacking strangers who have done nothing to them personally.

Zeynep Tufekci is a sociologist of the Internet who writes a blog called Technosociology. She is a fellow at the Center for Information Technology Policy at Princeton University and an assistant professor at the University of North Carolina. Her interests include the formation of social groups on the Internet.

Tufekci was familiar in general terms with the Amanda Knox case but not with the Internet furor. It didn't surprise her. "At first," she said, "it may look like an unstructured mob. But when you trace it back, there's a place where they coalesced. There's a community aspect to the swarming." She explained that it is the forming of the community that makes it formidable, effective, and long lasting. These are not anonymous crazies blogging alone in the dead of night. "They've invested in becoming a community." Without that community, she said, they probably wouldn't be able to sustain their activities.

But who are these people? While we were talking, Tufekci Googled around and noted the crude, sexual nature of much of the anti-Amanda commentary—from both male and female commentators. "This mob congregating on one person so viciously," she said, "could have a variety of motives. It might be a coalition of female-on-female competition and sexually interested males." They may have unexamined motives for wanting to see Amanda punished. "The Internet has given these punishers sudden access to a certain kind of power, and they are not fully connected to their victims. There is a disinhibition that comes from being online. This constant refrain that the Internet is just the Internet, that it's not real—this helps dissociate people from their actions. In people's minds, it gets construed as unreal space where you're not morally accountable. But of course this is nonsense. The Internet is very real. We're social animals. It has a powerful effect in the world." She suggested I look back in time and trace the development of the anti-Amanda blogosphere.

In the first few weeks after her arrest, the anti-Amanda comments seemed random and inchoate. But as time passed, a more organized movement developed. The anti-Amanda bloggers coalesced and created two websites, "Perugia Murder File" (which later split into two after a dispute, with extensions .net and .org) and "True Justice for Meredith Kercher."

Not all the bloggers at these sites appeared to fit the stereotype of losers who needed to get a life. Many appeared educated and intelligent. They wrote well. They were articulate. They were effective. They seemed to have friends and jobs. And they were utterly and completely obsessed. The chief moderator of "Perugia Murder File," Skeptical Bystander, according to statistics on her profile, has blogged about Amanda Knox an average of seven times per day, every day, for the past five years. The creator of "True Justice," Peter Quennell (the only anti-Amanda blogger to use his real name), wrote over eight hundred detailed articles about the case in addition to posting more than two thousand comments. His writings add up to more words than the Bible, *War and Peace*, *Finnegans Wake*, and the *Iliad* and *Odyssey* combined. Five and a half years later, all three of these websites are still going strong, especially after the March 26 verdict requiring that Knox be retried. These people, who saw their cause apparently lost with her acquittal in 2011, acquired a new lease on life with the verdict. They are ecstatic and have come roaring back, angrier than ever.

Almost from the beginning, the "True Justice" site became a clearinghouse of anti-Amanda fervor. It would eventually mushroom into a website receiving millions of hits. It specialized in long, detailed articles with bulleted points, footnotes, diagrams, and photographs, which dissected and evaluated every aspect of the case. There were analyses of the crime, the trial proceedings, the cast of characters, and the scientific evidence. There were gigabytes of PowerPoint presentations. The website opened files on people connected with the case, and it paid especial attention to Amanda's defenders, researching their backgrounds and raising questions about their motives, honesty, and qualifications.

The website wasn't entirely negative. It praised the integrity, incorruptibility, and perspicacity of Italian police and prosecutors. A particular hero of the site was Giuliano Mignini. The website's pages were garlanded with pictures of Meredith Kercher, and it featured articles about her life, condolences to her family, and expressions of mourning for the loss to the world by her death. It should be noted that Meredith Kercher was, by all accounts, a remarkable person, her death a terrible loss.

The tone of the site was one of measured outrage. The many articles with their masses of detail created a believable alternate reality. This reality painted a picture of Amanda Knox as a sexual predator, drug addict, and killer, whose beautiful face was a mask covering sexual depravity. She was the product of a dysfunctional and possibly incestuous family. In this alternate reality, her younger sisters (one was twelve) dressed provocatively and sexually, and they showed clear signs of psychopathology; if not placed in foster care they, too, might become killers. The "murderess" (the feminizing of the word was standard) was supported by a cast of "carpetbaggers"—that is, opportunistic lawyers, money-grubbing journalists (like me), glory-seeking FBI agents, corrupt judges, narcissistic criminologists, and unqualified forensic scientists, all of whom were "wading in the blood of a murdered girl" for fame and money. "True Justice" detailed how Amanda's family had hired an expensive, multi-tentacled PR firm, which had managed to mislead the national media, including the four national television networks and the *New York Times*. Dissenting posters at "True Justice" were banned and their opinions removed.

A person who knew nothing of the actual facts of the case might well have found the "True Justice" website to be informative, believable, and consistent. And there were many people out there who did not know the facts. "True Justice" over time would be consulted and used as a source by some major news organizations, most notably the BBC and *Newsweek/The Daily Beast*.

The online furor was not just white noise. It drove public opinion against Amanda. It influenced coverage by legitimate journalists. For example, Barbie Nadeau, a Rome-based correspondent who covered the case for *The Daily Beast*, wrote a book about the case, *Angel Face: The True Story of Student Killer Amanda Knox*. While the book included no footnotes or bibliography, it appears to have used information sourced from anonymous bloggers—identifiable as such because it was incorrect. Tina Brown, editor-in-chief of *Newsweek/The Daily Beast*, contributed the foreword to the book. In it, Brown wrote that "a merciless culture of sex, drugs, and alcohol" led to Amanda's "descent into evil," and she wondered if Amanda's "pretty face" was perhaps only a "mask, a duplicitous cover for a depraved soul." To see statements like these come from the pen of the editor-in-chief of *Newsweek* shows how deeply the noise of the blogosphere had penetrated legitimate journalism. Tina Brown was joined by other media personalities who appeared to have gotten much of their information from anti-Amanda online commentary. Ann Coulter wrote that, "Despite liberals' desperate need for Europeans to like them, the American media have enraged the entire nation of Italy with their bald-faced lies about a heinous murder in Perugia committed by a fresh-faced American girl, Amanda Knox."

For this piece, I interviewed as many of the dedicated anti-Amanda bloggers as I could get to correspond with me. I asked them what in particular had drawn them to the case. While I received staggeringly long replies, emails running to many thousands of words, not one was able to articulate the source of his or her passion, beyond general statements about victims' rights, wanting to see justice done, or seeking to protect our children from murderers. Skeptical Bystander told me she was originally drawn to the case because she saw a photo of Amanda Knox, mistook it for the victim, and thought, "gee, she looks more like a killer than a victim." Their level of self-awareness seemed in inverse proportion to their level of outrage.

The online furor against Amanda spilled over into the real world, dramatically confirmed by what happened to Steve Moore, a much-decorated former career agent with the FBI. Becoming a special agent at twenty-five, Moore had served as a counterterrorist specialist, certified sniper, and helicopter pilot. He participated in covert operations against the Aryan Nations and other white supremacist groups, and he ran the FBI unit responsible for investigating acts of terrorism against the United States in Asia. He retired from the FBI in 2008 and took a job as deputy director of security for Pepperdine University, in Malibu, California. A handsome, rumpled man, he was known for being funny, self-deprecating, blunt-talking, and extremely stubborn.

After retiring, he was bored. In late November 2009, his wife, Michelle, was watching a CBS News report about Amanda Knox and asked him to come over and take a look—that it seemed an innocent American girl was being railroaded in Italy for a murder she didn't commit.

"I dismissed it," Moore told me. "I told her that those people are invariably guilty." Michelle persisted. "Show me where this report's wrong."

Steve went online and started looking at the case. "Right away," he said, "I found serious, damning problems with each major piece of evidence.... The further I dug, the more it became obvious to me that it was absolutely a fabrication. Later, when I finally got hold of the crime-scene tapes, I realized it wasn't an accident: It was intentional. This was an *intentional* frame."

At first, Moore did nothing. Amanda and Raffaele's trial was almost over and the verdict would be announced in a few weeks. He was sure they would be acquitted. "In the U.S.," he said, "you don't get evidence into court unless it's totally unimpeachable. I thought Italy must be like the U.S. There was no evidence against Amanda."

On December 5, 2009, they were convicted of murder. "I had to go home from work I was so shocked," he said. "My ears were ringing. I realized I couldn't sit idly by." As a former FBI agent, he was in a position, he hoped, to do something. He went to the administration of Pepperdine and received verbal and written permission to advocate on Amanda's behalf.

Moore delved into the case, researching it in depth. When he was ready to go public, he made a splash.

On September 2, 2010, he appeared on three shows on the same day—the *Today* show with Ann Curry, *Good Morning America* with George Stephanopoulos, and the CBS *Early Show* with Harry Smith. He told Stephanopoulos that the evidence against Amanda was "ridiculous," Italian forensic techniques "horrible," and her interrogation "Third World." He said, "I am as certain of her innocence as I am of anything in the world."

Moore was devastatingly effective. The three main anti-Amanda websites went incandescent. "I'm used to people not liking me," Moore said. "I've had murderers threaten me. The Aryan Nations posted on their website that I was on their list, that I was an enemy to my race. I went up against Al Qaeda in Pakistan. But this was beyond belief. I'd never experienced anything like it. Even the murderers I put away didn't fight me as hard as these nutcases over at PMF and TJMK."

The posts attacked Moore, ridiculed and mocked his Christian faith, called him a pedophile, a white supremacist, a liar, and a fool. They accused him of molesting his daughters. His oldest daughter received menacing phone calls. Michelle, Steve's wife, became a particular object of attack. "Michelle was the recipient of some of the most vicious, nasty stuff, some of them saying 'I long to do to you what they did to Meredith, only it won't be as nice,' telling her she needed to be raped." Someone sent her pornography. The discussion about Moore on the three anti-Amanda websites ran to hundreds of posts. "True Justice" opened a separate file on Moore in its "Carpetbaggers" section.

Skeptical Bystander got the ball rolling at her website, PMF:

My bullshit detector is on high alert! ... When I was a kid, my mom would threaten to wash our mouths out with soap if we lied or swore. Next time Steve is on television, he'll probably have soap bubbles seeping from every orifice.

GEEZ Skep! Give a body a warning, would ya? I'd rather not even think about Skeevie, let alone his orifices. He's a walking, talking orifice.

he pretends to be on a crusade to save the Holy Land of Amanda, who's just another drunken immoral college kid like the ones he despises at Pepperdine, except she's a million times worse! Oh, but she's young, she's hot, she's got blue eyes, she's sexy, so he lost it. He's a real Jekyll-Hyde and so is his wife, she drove him to this insanity ...

Pepperdine immediately became aware of the online furor. Moore returned on September 3 to a "firestorm" at the university. Officials there had been following the posts at "Perugia Murder File," "True Justice," and elsewhere, and they were troubled. Ten days later, Moore got a letter from the administration, which demanded that he stop advocating for Amanda. Their reasons were, in part, that they were concerned about "the threat to the University's reputation as some begin to question your investigation, your qualifications as an expert in this matter, and your motives." The "some" could only refer to the blogosphere, as only anonymous bloggers had raised these questions. (The regular media had received Moore cordially and treated him as the expert he clearly was.)

"I was gobsmacked," Moore said. "Stunned. In shock. It wasn't just the vicious, malevolent, and defamatory criticism by PMF/TJMK... it was that some sophisticated people believed their garbage. That's the danger with these types of Internet trolls; if they use the right verbiage, callow people will believe it."

Moore felt too strongly about the Knox case to abandon it. Pepperdine fired him. The bloggers went wild with jubilation.

> There was no doubt that his suicide mission would fail... He is just another Knox agent in a long line of very ordinary sock puppets and he will not be the last one who will ruin him/herself for the murderess.
>
> Moore would seem to be the sacrificial moron... Go get him, Mignini
>
> I am glad another murderer sympathizer is getting what he deserves. He will not be the last one. More research is been done to make sure.

Moore sued Pepperdine for wrongful termination. Pepperdine settled for a "mutually satisfactory" sum. When I asked Moore what "mutually satisfactory" meant, he said, dryly, "I think I might say I was very satisfied with the settlement."

"I feel like I have a purpose in life now that I didn't have before," he said. "I see injustice out there. And who is better to help right injustice than someone who knows how the justice system works?"

The blogosphere didn't forget. For the past four years, Moore has been pilloried online. Even today, he and his family are contending with vicious, anonymous attacks. In November 2012, a blogger named BRMull went to Moore's daughter's website, copied some of her songs to "Perugia Murder File," and critiqued them. He appended a threatening message: *"yes Steve Moore, I'm talking about your daughter, BRMull plays for keeps."*

Moore said, incredulously, "And I haven't done anything to them!"

During this time, a different sort of affray took place at Wikipedia. In April 2010, a group of Wikipedia editors, some with high administrator status, replaced the rather thin article entitled "The Murder of Meredith Kercher" with a new one. The new article, dressed up in the usual objective language of the site, was, in the view of other Wikipedia editors, strongly biased against Amanda. But when these editors tried to add balancing information to the article, the anti-Amanda editors swiftly removed the edits, saying the material was based on unreliable sources. Among the sources they deemed unreliable were CBS News, U.S. Senator Maria Cantwell, criminologist Paul Ciolino, *Vanity Fair* reporter Judy Bachrach, and the Pulitzer Prize–winning *New York Times* columnist Timothy Egan. At least two of the Wikipedia editors posted diatribes against Amanda on anti-Amanda websites under their Wikipedia user names, calling into question their neutrality. One such editor was the same BRMull who threatened Steve Moore's daughter; BRMull was responsible for more than four hundred edits to the article and had opined on cafemom.com: "I am a well-known Knox hater and proud of it."

A fight started, with information being added and just as quickly removed. The anti-Amanda faction prevailed, blocking and banning at least eight editors. They also blocked the creation of a separate "Amanda Knox" entry in Wikipedia, with queries redirected to the Kercher article.

One of the banned Wikipedia editors, PhanuelB, posted an open letter and petition to Jimmy Wales, the founder of Wikipedia. "The Murder of Meredith Kercher article," the letter said, "in its present form is not written from a neutral point of view and bears little resemblance to what reliable sources have said about the case."

Not long after PhanuelB submitted the petition, Wales himself showed up at the "Talk" page of the "Murder of Meredith Kercher" article—the page in which editors can engage in freewheeling discussions. Wales had researched the history of the article and he weighed in decisively against the anti-Amanda edits. He particularly objected to the systematic exclusion of reliable sources. "*Is it true that people have been banned for completely neutral edits? Yes. Is it true that reliable sources have been systematically excluded? Yes. None of that is acceptable.*" He accused some of the editors of censorship, the gravest of Wikipedia crimes. An acrid online discussion followed, with some of the anti-Amanda editors attacking Wales himself. Wales finally wrote, exasperated, "*I am concerned that since I raised the issue, even I have been attacked as being something like a 'conspiracy theorist.' . . . Whenever we see outrage in the face of mere questions, it is good to wonder where the truth lies.*" Wales, who takes a democratic view of his organization, was loath to throw his weight around and block or unblock editors. The "Talk" page for the article ballooned as editors fought tooth and nail over every turn of phrase. The war over the article went on for months. Finally, around the time Knox was acquitted and the evidence against her revealed as bogus, the troubled article was junked and rewritten completely by one of Wikipedia's top editors, SlimVirgin, who produced a neutral, factual piece.

"These hard-core editors," PhanuelB told me, "were tough as nails. These were not stupid people. They had been involved in Wikipedia before us, which was why they defeated us initially." He noted that within twelve hours of his posting the letter, anonymous bloggers outed him.

> Want to see what this despicable man looks like? Meet PhanuelB otherwise known as Joseph W Bishop.

In addition to photographs, they posted many personal details, his place of employment, his home address, and even a photograph they found online of his family Bible. When they discovered that Bishop had served as an engineer in dangerous areas of Iraq, they went wild with speculation that his stint there had left him "mentally damaged."

Bloggers at PMF also went after Jimmy Wales, seeing a more sinister motive in his intercession than merely trying to maintain Wikipedia's neutrality. They latched on to a sexual controversy in his past and used it to claim he had a sexual interest in Knox.

> Right now, Jimbo is on the verge of losing any sense of respect as a new media entrepreneur ... just for a chance to catch a peek of some tender young sex killer flesh.

> Jimbo is precisely the profile of the aging Lothario looking for access to tail through his powerful media connections.

> Somebody should give Jimbo a cold shower. He's really lathered up and ready for brunette sex killer action

Skeptical Bystander added her own quip. "In all seriousness, what is it with these wiki guys and their wicks?"

The anti-Amanda group attacked many others who expressed pro-Amanda opinions, often by going after them at their places of employment. They did this in ingenious ways, sometimes by finding out who their supervisors were and attacking them, posting pictures and personal details of their lives, along with insults and threats. Such victims included a high school teacher in Hawaii, a professor at Leeds University, an employee at the Committee to Protect Journalists, a judge in Seattle, criminologists, attorneys, and scientists who did pro-bono work on behalf of Knox. The pro-Amanda bloggers fought back, often anonymously themselves, outing whenever possible the more active bloggers and turning nasty research back on them.

Peter Quennell was exposed for harassing a New York City ballet dancer and had a restraining order placed against him. BRMull turned out to be Brendan Robert Mull, a doctor in California who, they discovered, was on probation for attempting to strangle a female psychiatrist who had been treating him for drug and alcohol abuse. Pro-Amanda bloggers posted Skeptical Bystander's real name and personal information online, her husband's name, details of her life and shopping habits. They sent her emails threatening enough that she took them to the police, who conducted an investigation. Going after Amanda Knox turned out to be a risky business.

The fundamental question remains: Why did these many people, with no connection to the case and at potential risk to themselves, devote their lives to attacking Amanda Knox and all those who supported her? The answer to this baffling human behavior lies, as many such answers do, in evolutionary biology.

KATRIN RIEDL FROM the Max Planck Institute for Evolutionary Anthropology in Leipzig, Germany, performed a curious experiment with chimpanzees. She set up a situation where a chimpanzee, using a set of pulleys and traps, could steal food from another chimp. A third chimpanzee, observing the theft, could then "punish" the thief by pulling on a rope, depriving the thief of its ill-gotten food. The idea was to see if chimps engaged in "third-party punishment"—that is, if a chimp would punish another chimp for wronging a third chimp.

But the third chimp never punished the thief—not even when the victim was a close relative. This experiment and others showed that chimpanzees do not engage in third-party punishment. If a chimp steals food or commits a wrong against another, the victim will retaliate. But bystanders, even close relatives, will not intervene.

This is starkly different from human behavior. Other researchers at the Max Planck Institute did an experiment with three-year-old children. The experiment ran like this (I've simplified it a bit): A child was brought into a playroom, where there were two hand puppets of a cow and an elephant, manipulated by actors. Cow, Elephant, and the child then made their own sculptures out of clay. Cow made a flower, Elephant a snail, and the child created whatever she liked. Then Cow left the room. Elephant said to the child, "I don't like Cow's flower. I'm going to break it now." And Elephant destroyed Cow's sculpture and put it in the trash.

Almost all the children observing this protested, sometimes tried to intervene, and tattled when Cow returned. After that, the children were friendlier toward Cow, seeking to comfort it, patting and stroking, while turning a cold shoulder on Elephant.

The experiment and others showed that by three years of age, children already demonstrate a strong, innate, and sophisticated propensity to react to and punish third-party transgressions.

Some anthropologists call third-party punishment "altruistic punishment." Why "altruistic"? Because an individual who punishes a third party who has not harmed him directly, but has transgressed the norms of the group, does so altruistically—that is, for the good of the group with no personal gain. He also does so at personal risk, as the targeted individual may retaliate.

This behavior is unknown in chimpanzees. Which suggests that altruistic punishment is a unique product of human evolution.

One man who has spent the last decade studying the evolution of altruistic punishment is Samuel Bowles, former professor of economics at the University of Massachusetts and now research professor and director of the Behavioral Sciences Program at the Santa Fe Institute. Bowles has a Ph.D. in economics from Harvard. His research challenged the standard economic assumption that people are motivated entirely by self-interest, and this led him to study human evolution and the development of altruistic behavior.

Bowles has looked at this profound question by mathematically modeling the evolution of small human groups and comparing these models with studies of hunter-gatherer societies. He essentially asked the question: How did altruistic behavior evolve? Altruistic behavior on the surface would not appear to be adaptive. Someone who sacrifices or puts himself at risk for the benefit of the group isn't going to pass on his genes as readily as a selfish person who never sticks his neck out. So why aren't human societies made up entirely of selfish people acting in their strict self-interest?

The reason is group evolution. A group that cooperates, in which some individuals act for the benefit of the group, will prevail over a group of totally selfish people. But when Bowles mathematically modeled the evolutionary benefit of straight-up cooperation, he found it to be almost nonexistent. Groups of merely cooperative individuals don't evolve in a strongly cooperative direction, because of the problem of slackers. In such a group, it becomes everyone's best interest to be a freeloader—that is, a person who benefits from group cooperation without contributing. The slacker is the guy who sleeps under a bush while the rest go out hunting the mammoth, but then partakes in the feast afterward. To counteract freeloaders, the group needs punishers. It needs someone to say, *Hey, pal, you didn't hunt, you don't eat*. Bowles then modeled the evolutionary benefit of punishment to enforce cooperation. That had a powerful effect on the evolution of cooperation. Without punishment, cooperation in human society would not have evolved.

Here's how it works. Take a group of, say, two hundred people. Let's assume the group is composed almost entirely of cooperators, with a few slackers. The slackers do nothing, contribute nothing, but use up resources. They are a detriment to the group. If no one in this group is a punisher, the slackers get away with it and drag the group down. They weaken the group, make it less "fit" in evolutionary terms.

If you throw in a few "altruistic" punishers, a dramatic change happens. The freeloaders get punished. The number of slackers drops almost to nothing, the group benefits, and cooperative behavior evolves in a strongly positive direction. And as the number of slackers declines, the risk to the punishers for punishing also drops. The altruistic punishers have made the group stronger, better able to survive, and have done so at a diminishing risk to themselves.

Now assume another situation: Everyone in the group is a rabid punisher. Common sense tells us that this is a toxic situation and the group suffers. An ideal group, then, has a certain percentage of altruistic punishers in it. In such groups, cooperation evolves rapidly.

In other words, one of our most treasured human qualities—cooperation—evolved only because of the existence of punishment.

Bowles cites many psychology experiments that show human beings are avid to punish wrongdoing, even at expense to themselves. In one well-known study, college students were divided into pairs, A and B. A is given a hundred dollars and told that he can share as much or as little of that with B as he wishes. If B accepts the division, both get to keep the money. If B does not accept, both lose the money.

Logically, B would accept any amount of money from A—after all, it's free money. Not so. B will gladly accept half and will almost always accept forty dollars. But when A offers B, say, twenty dollars, B almost always refuses. Why? Because B wants to punish A for an unfair division, even though that also deprives B of money.

The experiment was extended. Now A shares money with B, with C as a witness. C has the option of punishing A if she thinks the division is unfair, but doing so costs C money. In chimp society, C wouldn't give a damn about A and B's sharing problem. But in human society, C will avidly punish A when the division starts to look "unfair," even at a cost to herself.

Other experiments showed that when people punish, the dorsal striatum, a reward part of the brain, lights up. Those subjects who sacrificed the most to punish got the biggest charge from it.

Bowles's mathematical simulations showed that an optimal society has a significant percentage of punishers. "There's quite a bit of evidence," he said, "that people really enjoy admonishing, inflicting harm on, and punishing other people who are breaking social norms. They *love* to punish." This, he points out, is a good thing. "A lot of the people who serve voluntarily in the military, or in criminal justice, are driven by motives of concern for other people. These are *good* people. If you look at history, what did liberal Europe create? It created a specialized group of people, wearing uniforms and badges, to enforce social norms" in a fair, evenhanded way.

As Bowles delved deeper into the mathematics of cooperation and punishment—particularly when he added warfare to the equation—a darker picture emerged, something he calls "parochial altruism."

If you run the same simulations, but now set groups against each other in warfare, with the weaker groups experiencing extinction, the mathematics run toward a situation like this: Within the group, slackers are dealt with harshly by punishers. Punishment is even more important, because a slacker in war can seriously endanger the group. Cooperation and altruism evolve even more strongly. "Groups with lots of altruists," said Bowles, "win wars." The losers die; they don't pass on their genes.

Let me pause to emphasize this disturbing point: Warfare in human history was essential for the evolution of cooperation and altruism.

"To call this controversial," said Bowles dryly, "is an understatement."

In the warfare scenario an additional, sinister effect becomes evident: Altruism within the group does not and must not extend to outside groups. The members of the other group must be demonized—otherwise, why would you kill them? They are bad to the bone and they must be punished. The altruism applies only within the group, not between groups. It is "parochial."

In our modern society, this sort of "parochial altruism" can—and does—go haywire. History is littered with examples: the witch trials of Europe, the Inquisition, pogroms, and innumerable irrational wars. In modern times Nazi Germany is a prime example. What was Kristallnacht but "altruistic punishers" gone mad, forming mobs that destroyed Jewish businesses and brutally murdered innocent people because they were the "other"?

Bowles said, "The sentiments that go into the most repugnant racial politics against outsiders come from the same evolutionary source that leads us to respond to natural disasters and helping others. This may be our legacy, that's how we got this way. But it doesn't have to be our destiny."

What does all this have to do with Amanda Knox? I will ask the reader to go back and read the comments quoted in this article—or better yet, Google your own assortment of nasty comments. You will find that most of them follow a similar pattern:

1. Amanda Knox has violated social norms (e.g., she is a sex pervert, murderess, liar).

2. An appropriate punishment is suggested (e.g., burning, raping, imprisonment for life, rotting in hell).

This is nothing more than third-party or "altruistic" punishment translated to the Internet.

Over the centuries most civilizations have developed state forms of social control and punishment. In our culture, for example, wrongs are investigated and evaluated by neutral parties (the "police"). A public airing of information ("evidence") is presented in an open, formalized manner with many people participating. The accused by law must be party to the discussions (the "trial"). The accuser must meet the accused face-to-face. All evidence, pro and con, must be considered. Every stage of the process is open and scrutinized. The state (not the victim's family) is charged with punishing, so there will be no feuds or vendettas. This process is one of the most precious assets of our society. It took Western civilization many bloody centuries to develop it.

Now we have the Internet. It functions, in part, as a non-state form of social control. But it is one where our punishing instincts go haywire. We earlier saw how anti-Amanda bloggers found one another, established websites, and became a community. We saw how these anti-Amanda communities waged an implacable cyberwar, in which anyone who expressed doubt about Amanda's guilt became the enemy and a target of the most brutal and irresponsible attacks. Employing a Mafia-like logic, they extended their attacks to friends, families, co-workers, and even the children of their targets. We saw how the desire to punish went absolutely berserk over Amanda Knox.

Never in human history has a system developed like the Internet, which allows for the free rein of our punishing instincts with no checks or balances, no moderation, and no accountability, and conducted with complete anonymity. On the Internet, any assertion, no matter how false, remains forever. It is a process that is horrendously unfair.

Community is a fundamental part of this process. The Internet simulates the small communities in which human beings thrive. But these Internet communities are devoid of the softening effects of real human interactions, in which discussions of wrongdoing occur face to face, where diverse opinions are expressed, and where people are held accountable for what they say. In these cybergroups, all are self-selected punishers. Dissenters are blocked and nonconforming opinions deleted. The accused is dehumanized. A toxic feedback loop of highly filtered information transforms the group into a cybermob not unlike the medieval witch hunts of Europe or lynch mobs in the American South. We see this phenomenon not just in the Knox case but all over the Internet.

The Internet is indeed a non-state form of social control—but one that is severely dysfunctional. The ugliness on the Internet is not white noise. It lasts forever. It cannot be ignored. It causes terrible things to happen in the real world. The Internet is a place where our darkest evolutionary biology runs riot.

UPDATE

In a way, I believe this is one of the most important articles I've written. The Internet has become one of the central pillars of our lives, and its dark underbelly has only gotten worse, a charnel house of anonymous trolling, abuse, lies, hate, conspiracy thinking, and cruelty that destroys lives, undermines democracy, incites toxic tribalism, and damages trust in America's fundamental institutions. Many people have tried to answer the question of why human beings behave this way online—anonymity, power, seeking meaning and community. But that doesn't answer the fundamental question: Why do people take pleasure in online trolling and abuse? The Internet's malevolency, this article suggests, lies in evolutionary biology, in particular the evolution of altruism, one of humanity's most wonderful qualities. But this also required the evolution of the desire to punish, the one quality not being able to evolve without the other, the two yoked together like Jekyll and Hyde. The Internet finally unleashed the Hyde in all of us, much to our sorrow.

Amanda was resoundingly acquitted by the Italian supreme court, which severely criticized police and prosecutors in the case, establishing her innocence beyond any doubt. After her return home, she became an activist, journalist, and author. She wrote a bestselling book about her ordeal, Waiting to Be Heard, *and became an advocate for the wrongfully accused. In many other ways she turned her grueling experience into doing good and helping others, even as she continues to be viciously hounded by anonymous trolls.*

OLD BONES

SKELETONS IN THE CLOSET

SOME YEARS AGO, I worked at the American Museum of Natural History in New York City—as a writer and editor, not as a curator. One morning I opened the door to my office and was nearly knocked down by the smell of mothballs. Later I complained to my co-workers at the coffee machine, and one of them suggested that I contact the anthropology department. "I think they've got some kind of storage room next to your office," he said.

Indeed they did. I learned over at the anthropology department that a wall of cheap plasterboard was all that separated me from the museum's collection of well-preserved human bodies. It seems that this particular morning the mummies had received a fresh change of paradichlorobenzene crystals to keep them free of insects. Curious, I decided to pay my neighbors a visit.

The mummies were stored in the defunct South American hall, a cavernous room with a tiled floor and fine old oak cabinets. Most of the mummies were stacked along the back wall in a solid tier of black tin crates; several in the center of the room were in glass cases—apparently they had once been on display.

It was that morning when I first began to understand that the American Museum collected not only the art and artifacts of other cultures but bodies, too, along with bones, skulls, whole skeletons—in a

---

Copyright © 1989 *Harper's Magazine*. All rights reserved. Reproduced from the February issue by special permission.

sense, collected people of other cultures. It is such a large collection that storing it is a headache. One curator had sacrificed half his office for the keeping of thousands of human skulls, each in its own little cardboard box. Lining the halls outside the anthropology department's offices were rows of lovely nineteenth-century cabinets; in many of them, behind rippled glass, I glimpsed stacks of human bones and mummified body parts. Nobody knew exactly how many individuals' remains (or parts of remains) the museum held, but my guess was close to 25,000—a very large graveyard.

My curiosity eventually led me to the museum's archives. What I wanted to know was, Where did all these remains come from? Why did the museum collect them? And what were they doing here now? You wouldn't know about them from visiting the museum's exhibition halls. It was as if they were a secret, a mystery.

Reading old museum reports, I learned that the stories of how the human remains got to the museum are in some cases as unsettling as the bones and skulls and mummies themselves. There is, for example, the story of the Fortress Rock mummies. In 1928, the museum launched the Stoll-McCracken Arctic expedition—actually a wealthy shooting party—to collect Pacific walrus in the Aleutian Islands for one of the museum's new habitat groups. (You can still see a few of the walrus brought back, now stuffed, in the Hall of Ocean Life.) But the expedition was after more than walrus. Anthropologists at the museum knew that in 1875 a sealer had unloaded in San Francisco a dozen mummies said to have been collected in the Aleutians.

The anthropologists, who were then deep into research on the origins of humans in the New World and relationships between the tribes, were interested in studying Aleut mummies and hoped more could be found; they sent an archaeologist named Edward Weyer Jr. along with the expedition for just this purpose. When the expedition's boat anchored at one or another Aleut port, Weyer made inquiries about old graveyards or deserted villages. During one stopover, he heard from several villagers about a "strange rock" in the Bering Sea, just north of Unalaska Island.

The spot was easily found: according to Weyer's writings, the members of the expedition saw a "great abrupt rock, which was cleft near its landward end by a deep precipitous gorge," rising from the sea. They named it Fortress Rock, because it resembled a grim medieval castle. Weyer and his assistant landed on the island's shingle beach and scaled the cliff with ropes and axes.

The island turned out to be a kind of Aleut mausoleum. Most of the summit was covered with shallow graves, and a quick search of the caves along the cliffs yielded dozens of skulls. At one end of the island, Weyer's assistant discovered a buried crypt constructed of expertly mortised driftwood timbers and secured with ivory nails. The crypt had been caulked and lined with sealskins and woven grass. Inside he found exquisite rolls of bird skins sewn together, ivory harpoons, stone lamps, beads of amber, and other offerings—as well as four tightly lashed and wrapped bundles. Inside each bundle was a well-preserved human being: two men, a woman, and a child.

Weyer and his assistant lowered the artifacts and mummies on litters to the base of the cliff, where they were packed in crates and shipped to New York. The museum considered Fortress Rock a major discovery.

The Fortress Rock mummies have never been placed on display by the museum. It is possible that they will remain forever in one or another of the museum's storage rooms, given a fresh change of moth flakes every now and then. But I doubt it. In the past few years a small problem has come up. Mummies, as well as thousands of other remains in museums, have become an issue.

Across the country, Native American tribes—by Native American I mean American Indian, Inuit, Aleut, and native Hawaiian—and Pan-Indian groups, such as the Native American Rights Fund and the National Congress of American Indians, are demanding that mummies and skeletal parts held by museums be returned to them for reburial.

The American Museum of Natural History is by no means alone in housing large collections of Native American mummies and skeletal parts. The Smithsonian's National Museum of Natural History has about 18,500 specimens (each "specimen" might be anything from a few bones to a complete skeleton or mummy); there are perhaps another 5,000 at Harvard's Peabody Museum; the National Park Service has perhaps as many as 20,000 specimens tucked away in repositories all over the country. The Native American Rights Fund estimates that there might be as many as 600,000 such specimens in museums, historical societies, universities, and private collections in the United States.

Several state legislatures (Alaska among them, at the request of the Eskimo and Aleuts) have passed resolutions calling for the return of all Native American specimens held by the Smithsonian. Legislation on the issue is slated to be introduced in Congress this year. [This important legislation, called the Native American Graves Protection and Repatriation Act, or NAGPRA, passed in 1990 and transformed American archaeology.] If it passes, the federal government would begin to assist Native Americans who want remains returned to them. And if Native Americans are able to get the bill they want passed, the new law would force the museums to yield their collections. Already, without federal action, numerous tribes (with the aid of activist groups) are moving ahead on their own—pressuring museums, even threatening to sue, for the return of remains.

To many Native Americans, the collecting of their ancestors' bones and bodies by museums is a source of pain and humiliation—the last stage of a conquest that had already robbed them of their lands and destroyed their way of life. "They took everything," Walter Echo-Hawk, a Pawnee and staff attorney with the Native American Rights Fund, said to me recently. "Including our dead. Even our dead." Native Americans argue that museums have had decades to study these bones. They also wonder why museums *need* thousands of skeletons. To them, the scientific interest in Native American remains smacks of racism, as if they were freaks or curios. "Let them study Germans or Swedes for a change," one Aleut I spoke to told me.

Physical anthropologists I talked with are aghast at the possibility that they might have to surrender their collections of Native American remains—what they call their "database." They explained that there have been important developments in the past decade, the discovery of new techniques for analyzing bone and desiccated tissue. In the next ten years, I was repeatedly told, the careful study of these bones could well yield answers to some of the deepest questions in American anthropology—including questions about the very nature of the conquest of the Native Americans. Museum administrators, too, are anxious about, and at times baffled by, the desire on the part of Native Americans to retrieve and rebury their ancestors' remains. The issue, in a sense, attacks the museums at their heart; the perpetual care of the collections, in light of the Native Americans' demands, begins to seem a barbarous act. Entire worldviews can appear at times to be butting against each other: for a curator or researcher, to rebury something is to destroy it. Natural history museums exist primarily to hold things, not only for current research but—most importantly—for whatever research might be conducted in the future. Because no one can predict what that future research might be, virtually nothing can be thrown away.

However, for some of the Native Americans I spoke with, this approach to scholarship and research is nothing if not otherworldly. How, say, could the needs of science compare with the fact that their grandfathers' spirits are forced to wander unceasingly because their bones are in a box at the Smithsonian?

In the spring of 1986, a group of Northern Cheyenne chiefs went to Washington at the invitation of Senator John Melcher, a conservative Montana Democrat. The purpose of the visit was to try to recover one of the tribe's sacred Sun Dance songs that had been lost many years before, but which they hoped might have been recorded on wax cylinders now stored at the Library of Congress. The chiefs quickly succeeded in recovering a lost song, and in the time remaining, they decided to go on over to the Mall to have a look at the Smithsonian's Cheyenne collection. They spent an afternoon in a large storage room on the top floor of the National Museum of Natural History, poring over photographs and accession records.

Last fall I spoke to Clara Spotted Elk, a Northern Cheyenne Indian and legislative assistant to Senator Melcher, who had made the arrangements for the Cheyenne to travel to Washington and search for their songs. "As we were walking out," she said, "we saw there were huge ceilings in the room, with row upon row of drawers. Someone remarked that there must be a lot of Indian stuff in those drawers. Quite casually, a curator with us said, 'Oh, this is where we keep the skeletal remains,' and he told us how many—18,500. Everyone was shocked. I mean, it was such a shocking thing that no one said anything. The chiefs were quite alarmed because we had been sitting there all day with those restless spirits. So we really beat it out of there.

"A few days later, I related this incident to Senator Melcher. He said, 'Young lady, you've got to learn to get your facts straight. The Smithsonian couldn't *possibly* have 18,500 Indian skeletons rattling around in the attic.' So I checked into it and reported back that, yes, indeed, they had 18,500 skeletons. He was outraged." Shortly thereafter, Melcher and his staff began to draft the Native American Museum Claims Commission Act, known in Washington and among Native American activists as the "Bones Bill." The bill was approved in committee but got no further in the 100th Congress. Melcher lost his Senate seat last November [1989], but it is expected that Democratic Senator Daniel Inouye of Hawaii will reintroduce the bill this year. Many museum officials and Indian activists alike believe that the Bones Bill in one form or another will soon be passed.

The Bones Bill, as drafted by Melcher, would apply to most Native American remains, as well as "grave goods" and religious artifacts in public American collections. While the details remain to be debated and hashed out, here's how such a bill would probably work: If a tribe can show that a collection of remains is either clearly from its tribe or had been dug up from its ancestral burial grounds, then the tribe can request that the remains be handed over to it. The only way a museum could keep the remains is to show proof that the bones had been dug up with tribal permission. The bill would establish a national commission to mediate disputes between tribes and museums.

The bill promises to raise all kinds of complex legal questions. Are remains "abandoned property" or do they belong to the descendants, even if those descendants are no longer aware of them? And how does one define "descendant"? The Smithsonian has large holdings of tribal remains originally picked up by Army doctors and curio collectors on battlefields. Who has a right to these? And what about extinct cultures? Can Pan-Indian groups legitimately claim (as some do) that they speak for the dead of a thousand years ago?

Even without the Bones Bill, Native Americans have been active in seeking the return of Indian remains. In some cases tribes have simply asked museums for remains, but increasingly they have raised the possibility of lawsuits. The Smithsonian has been approached by the Aleuts of Kodiak Island, Alaska; the Oglala Sioux of South Dakota; and fourteen other tribes. Native Americans have also demanded the return of skeletons held by the National Park Service, the Army Corps of Engineers, and some smaller museums and historical societies in the West. "And this," said Christopher Quale, an attorney for the Three Affiliated Tribes of the Fort Berthold Reservation, in North Dakota, "is just the beginning. It's conceivable that some time in the not-so-distant future there won't be a single Indian skeleton in any museum in the country. We're going to put them"—he meant the physical anthropologists—"out of business."

THE GROWING BATTLE between Native Americans and American museums affords its share of ironies. For the first time, the traditional defenders of Native American rights and culture—the anthropologists and the museums—suddenly find themselves and their values under attack by the very people they have devoted themselves to observing, researching, caring about. Between 1880 and 1930, when American Indian society and culture were being decimated, the anthropologists and the museums were the only forces in our society working to protect—or at least to save—what remained of this heritage. As a result, collecting institutions ended up preserving many aspects of Indian culture (such as the Cheyenne Sun Dance song) that otherwise would have disappeared.

Ironic too is the fact that during the past twenty years, while many Native Americans have sought to revive their traditions, it is to the museums and anthropologists that they have turned. The museums have the photographs showing how things once looked, the descriptions of sacred rituals, the sacred artifacts and songs. They captured for posterity many things that were forgotten or lost during the time when the government was actively suppressing Indian culture. Originally, the museums and the anthropologists welcomed the Native American "revival." But now?

Another irony: Native Americans are receiving some of their strongest support from the traditional opponents of the Native American rights movement—the white, Western conservatives. In California, for example, right-wing Christian fundamentalists have been among the Indians' most effective allies. Western politicians in chronic trouble with the Indian tribes (as was Senator Melcher) can champion the Bones Bill without alienating their white conservative support—it's not expensive "welfare" legislation, nor does it involve relinquishing any land. Moreover, the issue can be used (and has been) to bash the liberal, elitist Eastern museum and scientific establishment—always a crowd pleaser.

Most museum officials and anthropologists realize it will be very difficult for politicians to oppose the return of remains, and for this reason, they are very worried. The American Association of Museums has been working on a reburial policy of its own—one, naturally, that would allow museums to hold on to more remains than they might under more stringent legislation. "If we don't do it," Edward H. Able Jr., executive director of the association, told me, "then someone else will do it for us." In all my years of working in and around museum people, I have never seen anything like the uneasiness this issue inspires. One curator, on hearing the subject of my phone call, blurted out, "Oh my God." And an eminent physical anthropologist, after a long and unresponsive interview in which he repeatedly denied there was a problem, suddenly broke off and said: "Why are you doing this to us?"

IN THE PENDING battle over collections of Indian remains, two questions will no doubt be frequently asked: Why were these bones collected in the first place? And what scientific purpose do they serve today?

In 1886 a young German anthropologist named Franz Boas arrived in the United States with a radical idea, one that would become the cornerstone of modern thinking about race and culture. This was the idea of cultural relativism. Boas argued that human races were *intrinsically* equal—equally advanced and equally expressive of the complexities of the human spirit. Thus, the smallest tribe with a distinct culture was as important to anthropology as the great edifice of Western civilization itself.

Boas's views took hold in the lecture hall and then in society at large. Cultural relativists saw that the fantastic diversity of human cultures would soon be gone, swept away by war and progress. And thus began a frenzied period of collecting that would last half a century. If cultures could be "saved," they should be. But if the culture didn't survive, the new experts felt that science could at least assemble a complete record of it, a mass of raw data for future study. Photos were taken, plaster casts were made. And the researchers dug in native gravesites, gathered skeletons and mummies, and carried them off.

The people whose ancestors were being "collected" were never seriously consulted; no one worried much about their beliefs, values, feelings. Boas himself, in the dead of night, raided the graveyards of his beloved Kwakiutl tribe of British Columbia. "It is most unpleasant to steal bones from a grave," he later wrote in his diary, "but what is the use, someone has to do it." The values of science, Boas would say again and again, were supreme: it was a matter of cultural history, not Kwakiutl mores. Or even Christian mores, for that matter: museums are holding large collections of the remains of other races (particularly whites) as well.

Museums continue to defend their early collecting practices. "Our collections were gathered according to the legal and ethical standards of the time," one museum director told me. "There's no way you can go back and rejudge the past." But even by the standards of the day, the methods of collecting were deemed sufficiently shocking to be kept out of the public eye. Consider the case of the six Inuit who came to the American Museum of Natural History in 1896. The Arctic explorer Robert E. Peary had brought them from Smith Sound, Greenland, to New York City—he had reportedly asked members of the tribe if any of them wanted to "visit" New York. As the Inuit lived closer to the North Pole than humans anywhere else on the earth, Franz Boas and his colleague Aleš Hrdlička were particularly eager to study them, and offered them spacious accommodations on the museum's fifth floor. On October 4, the Inuit tour group arrived. It was noted in passing that they all had "slight colds." Hrdlička immediately began measuring and photographing them. He also made casts of their faces, arms, and legs.

In four of the six Inuit, the slight colds developed into tuberculosis, to which they had no natural immunity. By spring, despite everything that could be done, all four had died. Hrdlička and Boas quickly went to work. Here was a splendid, unparalleled opportunity to add postmortem data to their Inuit file. Hrdlička directed that all four be macerated, boiled, and reduced to skeletons at the College of Physicians and Surgeons of Columbia University. He then installed the skeletons in the museum's collection, where he could study them at leisure.

WHY DO MUSEUMS want all this now? Do mummies and bones serve any useful scientific purpose today?

Physical anthropologists used to think that by measuring bones they could answer important questions about the origin and spread of human beings across the New World and the exact relationships between the various races. They measured the bones of thousands of skeletons and tried to quantify physical similarities and differences. Despite tomes packed with tables, graphs, and charts, early physical anthropology pretty much failed to answer the bigger questions; the work yielded only empty classifications of "physical types." Physical anthropology as it applied to the modern races gradually became unfashionable and the skeletons languished unstudied, for the most part, in museum drawers.

But in recent years these remains have apparently become valuable once more to researchers. I spoke with Douglas Owsley, an associate curator at the Smithsonian, who is one of the staunchest defenders of the museum's collection of human remains. In just the past few years, he said, biomedical researchers have been developing techniques to extract certain proteins from human bones—proteins called immunoglobulins that are generated to fight off disease, and that remain in the bones in trace amounts long after death. According to Owsley, the bones in museums today might soon be able to tell us about the diseases people once contracted. "This technique," Owsley said, "will help us track the history of human diseases, the antiquity of diseases, even the evolution of diseases." Not only does this promise to revolutionize our understanding of the past, he added, but it might prove a powerful new tool for fighting disease in our time.

Owsley also mentioned that human DNA can now be isolated from mummified tissues, and molecular biologists at Berkeley are working to isolate DNA from archaeologically retrieved bones. (DNA has already been extracted from an 8,000-year-old preserved human brain.) If this technique works, Owsley said, "We will be able to study directly the genetics of ancient populations. From that, we can reconstruct such things as the travel movements of ancient peoples."

You may ask (as many Native Americans do): Couldn't museums just keep a dozen or so skeletons from each tribe, and give the rest back? This is not, alas, the way science works. To arrive at general truths, to extrapolate from the particular to the general, scientists need to study large numbers of objects. Scientists testing new drugs need large sample groups in order to say with any certainty whether something works. In a sense, the same thing holds true for physical anthropology; and the more powerful and refined the techniques become, the more important it is to work with large numbers of remains.

David Hurst Thomas, curator of anthropology at the American Museum of Natural History, is a specialist in Native American prehistory. He sometimes excavates human remains, and he has had Native Americans try to shut down his sites. Thomas has done a lot of thinking about the issue.

"The body," he said, "carries forward a record of almost everything that happened to that person in life." Episodes of starvation and disease, he explained, leave marks on bones, much like tree rings.

I asked Thomas to give me some specific examples of research that could be done using bones. "Well," he said, leaning back in his chair, "should Father Junípero Serra, the founder of the Spanish missions in California, be made a saint?"

Many American Catholics say yes, that Serra brought Christianity and a better way of life to thousands of Indians. The Catholic Church has cautiously agreed, and has advanced Father Serra to beatification, one rung below sainthood. California tribes have angrily protested, arguing that Father Serra's missions were little better than concentration camps where brutal slave labor, starvation, and disease killed all but a fraction of the native population.

"So how do we resolve this?" Thomas continued. "Well, you go to those old missions, excavate the Indian remains, and see what people were dying of." Archaeologists in California proposed to do just that. Their effort was blocked by tribes there, who did not want the remains disturbed. "It's too bad," said Thomas. "An examination of those bones could document precisely what the Indians have argued, that they were badly mistreated by the missionaries during that period."

Thomas himself is sympathetic to Native American claims and feels that many "historic-period" skeletons—those of Native Americans thought to have lived after the arrival of the Europeans—ought to be reburied. He will not excavate a grave without getting the permission of the group he feels is most closely related to those whose remains he plans to dig up. He will eventually rebury all the historic-period Indian skeletons he excavates, but he plans to keep track of where each one is reburied—a process he calls "in-the-ground curation"—so that anthropologists in the future can easily (but respectfully, he emphasized) locate and reexamine a specific skeleton.

"In many ways," he said, "anthropologists and archaeologists brought this whole situation upon themselves. There are those who think that this is just a political flap that will blow over, and then they can go back to the good old days when they could pop a burial whenever they wanted. They are sadly mistaken. Unless museums willingly respond to these concerns, we will be put right out of business."

I WAS CURIOUS to see how a typical reburial request was coming along, so I contacted members of the Oglala Sioux tribe in South Dakota. On June 15, 1988, the tribe had sent a short letter to Robert McCormick Adams, secretary of the Smithsonian, asking for all Oglala Sioux remains held by the Smithsonian. These, according to the Oglala Sioux, number 350 to 400 specimens, including the remains of three known individuals: Smoke, Two Face, and Black Feet.

The Sioux, and the Oglala Sioux in particular, are among the tribes that resisted most fiercely the white encroachment on their land. At one time the Sioux, a nomadic tribe, ranged over an area encompassing the northern Great Plains and Western prairies, from Wisconsin to the Dakotas. Today the Oglala Sioux occupy a two-million-acre reservation in Pine Ridge, South Dakota. The tribe suffers an eighty percent unemployment rate; alcoholism is a commonplace; and the suicide rate is several times the national average—all a result, anthropologists and many Native Americans alike will tell you, of the cultural despair that followed the destruction of their traditional way of life. Tribal leaders believe that one way to alleviate these woes is to try to recapture, as much as possible, the old traditions. And they see the reburial of their dead as an element in this effort.

I spoke with Severt Young Bear, a fifty-four-year-old Oglala Sioux deeply involved in the reburial issue. In a slow, quavering voice, barely audible over a crackling long-distance line, he talked about why they wanted the remains back. "In Lakota"—the Indian name of the Sioux subgroup that includes the Oglala—"we have a red road we walk on in this earth. Lakota view the spirit of a person as being entirely different from the Christian view. In Lakota, after death we take care of the spirit. If you disturb that spirit it starts wandering. The spirit of my grandfather Smoke is still walking back and forth from his [burial] hill to Washington."

Severt Young Bear said that the tribe might have a traditional recalling-of-the-spirit ceremony between Washington and Pine Ridge. "We make four stops and ask the spirit to return back home, so the spirit won't be lost in the archives of the Smithsonian."

He talked about the tribe's plans for reburial. "We want to bury them in May. It's beautiful here in May. The grass is green and flowers are growing everywhere. That would be a good time." He also talked about Sioux history, about how the Sioux were once the greatest warriors in North America. Severt Young Bear worried that the tribe had no money to cover the transportation and burial costs and would try to find funding, somewhere. He was concerned about some of the young men of the tribe, who wanted to go get the remains right away and bring them back in a U-Haul. He felt this would be undignified and disturbing to the spirits.

Will there be any problem, I asked, with the Smithsonian? Have they been cooperative?

He seemed surprised by the question. "We haven't seen anybody opposing it. Why would they?"

I then called the Washington law firm of Hobbs, Straus, Dean & Wilder, which I understood to be the tribe's general counsel. There, Karen Funk, a legislative analyst involved in the matter, quickly dismissed any notion I might have developed that the Oglala would be getting their remains back anytime soon.

"Except for those identified by name," she said, "I'm skeptical that they can get any of those remains back at all. The Smithsonian has asked the tribe to prove that it, the museum, doesn't own those remains. With the Smithsonian's admittedly poor records, I can't see how the tribe will be able to prove anything." She explained that although the firm had helped the Oglala initially, it now feels that any further legal pursuit would be far too costly. The tribe must simply pressure the Smithsonian privately, on its own.

A NUMBER OF anthropologists I spoke with pointed out that once a favorite pastime of many tribes was the desecration of the graves of their enemies (which is true); others noted that tribal burial customs and ceremonies have changed (also true). They said that the tribes of today simply don't have the same culture, ceremonies, and beliefs as the tribes of the past, and therefore do not represent them. They objected most vehemently to the Pan-Indian movement's view that *all* Native American remains are sacred, even those taken from thousand-year-old burial grounds.

To my mind, these objections miss the point. Anthropologists, under other circumstances, will be the first to tell you that cultures evolve. Moreover, it was the white man who forced Indians to adopt a new way of life—*our* way of life. The real issue, it seems to me, is that most Native Americans feel deeply about reburial, for whatever reason. It is not for us to judge the legitimacy of this feeling. "All we're asking for is a little common decency," said Walter Echo-Hawk of the Native American Rights Fund. "We're not asking for anything but to bury our dead." It is as simple, and as complicated, as that.

UPDATE

The Native American Graves Protection and Repatriation Act was passed by Congress in 1990, after this article was published. While many museums and scientists initially complained about the law and claimed it would hinder free and open scientific research, over the years it has been broadly accepted. In many ways NAGPRA profoundly transformed American archaeology, forcing archaeologists to collaborate with Native American tribes and take into consideration their customs, sensitivities, and religious beliefs. It also forced hundreds of museums and historical societies to return ill-gotten human remains and grave goods to the tribes they once belonged to. NAGPRA has been invoked by the FBI to repatriate sacred objects from auction houses and galleries to the communities from which they were stolen.

CANNIBALS OF THE CANYON

MORE THAN A century ago, American travelers in the Southwest were astounded to find ruined cities and vast cliff dwellings dotting the desert landscape. Surely, they thought, a great civilization had once flourished here. It looked to them as if the people who created it had simply walked away and vanished: the ruins were often littered with gorgeous painted pottery and also contained grinding stones, baskets, sandals hanging on pegs, and granaries full of corn. The Navajo Indians, who were occupying much of the territory where this lost civilization once existed, called them the Anasazi—a word meaning "Ancient Enemy"—and they avoided the ruins, believing they were inhabited by *chindi*, or ghosts.

Not surprisingly, American archaeologists focused on the Anasazi and their great works, and they became the most intensely studied prehistoric culture in North America. A standard picture emerged, based on wide-ranging excavations of sites and on detailed ethnographic research among the Hopi, Zuni, and other Pueblo Indian tribes, who are the Anasazi's descendants. The Anasazi were—so the findings suggested—peaceful farmers, and they attained astonishing results in engineering, architecture, and art. The center of this cultural flowering, from the tenth century to the twelfth, seems to have been Chaco Canyon, New Mexico, a windswept gulch in the apparently endless sagebrush desert of the San Juan Basin. Chaco is marked by

---

Originally published in the *New Yorker* in 1998.

immense stone structures, some up to four stories high, called Great Houses. The largest, Pueblo Bonito, contains some six hundred and fifty rooms, and its construction required more than thirty thousand tons of shaped sandstone blocks. The Chaco Anasazi also built hundreds of miles of roads that stretched out from Chaco Canyon across the landscape in arrow-straight lines—an engineering marvel achieved without compass, wheel, or beast of burden. They erected shrines, solar and astronomical observatories, irrigation systems, and a network of signaling stations. They constructed more than a hundred Great Houses outside Chaco Canyon, spreading them over fifty thousand square miles of the Four Corners region of the Southwest. Many of these outlying Great Houses seem to have been connected to Chaco by the radiating pattern of roads. Archaeologists today call this cultural explosion "the Chaco phenomenon." But the phenomenon ended abruptly around 1150 A.D., when a vast collapse apparently occurred, and Chaco, along with some of the outlying sites, was largely abandoned.

Equally remarkable was the Chaco society. It seemed to be almost utopian. The Anasazi, the traditional view held, had no absolute rulers, or even a ruling class, but governed themselves through consensus, as the Pueblo Indians do today. They were a society without rich or poor. Warfare and violence were rare, or perhaps unknown. The Anasazi were believed to be profoundly spiritual, and to live in harmony with nature.

As a result, the Anasazi captured the fancy of people outside the archaeological profession, and particularly those in the New Age movement, many of whom see themselves as the Anasazi's spiritual descendants. The ruins of Chaco Canyon have long been a New Age mecca, to the point where one of the sites had to be closed, because New Agers were burying crystals and illegally arranging to have their ashes scattered there. During Harmonic Convergence, in 1987, thousands gathered in Chaco Canyon and joined hands, chanting and praying. People have also flocked to the villages of the present-day Pueblo Indians—the Hopi in particular—seeking a spirituality outside Western civilization. The Hopi themselves, along with other Pueblo Indian descendants of the Anasazi, feel a deep reverence for their prehistoric ancestors.

In 1967, a young physical anthropologist named Christy Turner II began looking at the Anasazi in a new light. He happened to be examining Anasazi teeth in the Museum of Northern Arizona, in Flagstaff, attempting to trace a peculiar trait known as the three-rooted first molar. On the last day of his research, he asked the curator to pull down a large, coffin-shaped cardboard box from a top shelf. The accession record said that the box contained remains from a remote area along Polacca Wash, an arroyo situated below First Mesa, on the Hopi Indian Reservation. The remains had been excavated in 1964 by an archaeologist named Alan P. Olson. Turner removed the lid and found himself gazing at a bizarre collection of more than a thousand human bone shards. Thirty years later, when he described the experience to me, the memory was still vivid. "Holy smokes!" he recalled having exclaimed to himself. "What happened here? This looks exactly like food trash." The fragments reminded him of broken and burned animal bones that he had found in prehistoric Anasazi garbage mounds. As he looked more closely, another thought struck him. Like many physical anthropologists, he had sometimes done forensic work for police departments. Once, in California, the police had asked him to examine some remains that had been found in the Oakland Hills—a skeleton still wearing a pair of boots. Turner had informed the police that the person had been savagely beaten to death. "Now," he told me, "I could see the same violence done to the Polacca Wash bones."

Turner borrowed the bones from the museum and took them to Arizona State University, where he was a professor. In 1969, he presented his findings in a paper he read at an archaeological meeting in Santa Fe. Word of what the paper would say had got around, and the room was packed. Co-written with a colleague of Turner's named Nancy Morris, the paper was entitled "A Massacre at Hopi." Turner informed the audience that the bones belonged to a group of thirty people—mostly women and children—who had been "killed, crudely dismembered, violently mutilated," and that the heads, in particular, showed extreme trauma: "Every skull is smashed, chiefly from the front, and massively so.... The faces were crushed while still covered with flesh." Most of the skulls had received a number of "blunt, heavy, club-like fracturing blows." The bone material had still been "vital" at the time the blows were struck. "The many small pieces of unweathered teeth and skulls and postcranial scrap suggest, but do not prove, that the death of these people occurred at the burial site." Moreover, "every skull, regardless of age or sex, had the brain exposed." Heads had been placed on flat rocks and smashed open, apparently so that the brain could be removed.

He went on to say that most of the bones—not only the skulls—also showed marks of cutting, chopping, dismemberment, butchering, "defleshing," and roasting. The larger bones had been broken apart and the marrow scraped out, or, in the case of spongy bone, reamed out. Turner and Morris concluded that the Polacca Wash bones represented "the most convincing evidence of cannibalism in all Southwest archaeology."

Turner said that Olson, the original excavator, had been wrong in assuming that the bones were prehistoric, for Turner had had the bones radiocarbon dated, and the results had come back as 1580 A.D. plus or minus ninety-five years. Given the date, Turner wondered whether some record or legend might still exist relating what happened.

When he talked with me, he said, "We knew who the bods were. There were a certain number of kids and females. We looked at dental morphology. We got a good match with Hopi. So we asked ourselves, 'What is there in oral tradition about a whole bunch of Hopi being killed and eaten, or massacred, of this age and sex composition, at this date?'"

Turner eventually concluded that the Polacca Wash site was a place known in Hopi legend as the Death Mound. Hopi informants had first described the legend to an anthropologist at the end of the nineteenth century. According to the story, sometime in the late 1600s a Hopi village called Awatovi had been largely converted to Christianity under the influence of Spanish friars. In addition, the people of Awatovi practiced witchcraft, which the Hopi considered a heinous crime. Eventually, five other Hopi villages decided to purge the tribe of this spiritual stain. An attack was organized by the chief of Awatovi himself, who had become disgusted with his own people. Warriors from the other villages attacked the errant village at dawn, surprising most of the men inside the kivas—sunken ceremonial chambers of the Pueblo Indians—and burning them alive. After killing the men, the warriors captured groups of women and children. As one of these groups was being marched away, a dispute broke out over which village would get to keep the captives. The argument got out of hand. In a rage, the warriors settled it by torturing, killing, and dismembering all the captives. Their bodies were left at a place called Mas-teo'-mo, or Death Mound. "If the stories are correct," the anthropologist who first collected these legends wrote, "the final butchery at Mas-teo'-mo must have been horrible."

Turner recalls that the lecture room was quiet when he finished arguing that the bones were the remains of a cannibal feast. "You could *smell* the disbelief," he said. Most of his colleagues felt that there simply had to be another explanation for the strange bone assemblage. To suggest that the Hopi could have deliberately tortured, murdered, mutilated, cooked, and eaten a defenseless group of women and children from their own tribe seemed to make a mockery of a hundred years of cautious, diligent scholarship. The paper was looked upon with deep skepticism by many of Turner's peers, and the Hopi objected to what they considered a crude slur on their ancestors.

Over the next thirty years, Turner looked deeper into the archaeological record for signs of cannibalism—going all the way back to the Hopi's Anasazi ancestors. To his surprise, he discovered that a number of claims of Anasazi violence and cannibalism had been published by archaeologists, but the profession, perhaps blinded by the conventional wisdom, had ignored the reports, the notes, the evidence, the very bones.

Turner has identified many Anasazi sites that he believes represent "charnel deposits"—heaps of cannibalized remains. Next month, the University of Utah Press will publish the results of his work, under the title *Man Corn: Cannibalism and Violence in the Prehistoric American Southwest* [1998]. The term "man corn" is the literal translation of the Nahuatl (Aztec) word *tlacatlaolli*, which refers to a "sacred meal of sacrificed human meat, cooked with corn." (The book had a co-author, in Turner's wife, Jacqueline—also an anthropologist—who died in 1996.)

Man Corn reexamines seventy-two Anasazi sites at which violence or cannibalism may have occurred. Turner claims that cannibalism probably took place at thirty-eight of these sites, and that extreme violence and mutilation occurred at most of the others. He calculates that at least two hundred and eighty-six individuals were butchered, cooked, and eaten, with a mean of between seven and eight individuals per site. As a test to see how widespread cannibalism might have been, Turner also examined a collection of 870 Anasazi skeletons in the Museum of Northern Arizona. He found that eight percent—one skeleton in twelve—showed clear evidence of having been cannibalized.

Turner's book does more than attack the traditional views of the Anasazi. It also addresses one of the great unsolved problems of American archaeology: What caused the collapse of the Chaco culture around 1150? There was a severe drought at the time, but most researchers don't believe that it alone could have brought about such a cultural implosion. Other, unknown factors must have played a role.

After the Chaco collapse and abandonment, many Anasazi moved into deep, remote canyons, building their dwellings in cliffs or on high, often fortified, mesas. A century later, they abandoned even these defensive positions, leaving almost the entire Four Corners region uninhabited. To some archaeologists, it seemed as if the Anasazi had been seized with paranoia—that they were protecting themselves from some terrible enemy. But, no matter how hard archaeologists looked, they could find no such enemy.

I HAVE BEEN following Turner's work for about ten years. He is one of the country's leading physical anthropologists and has a reputation for being something of a loner—brilliant, arrogant, even intimidating. When I heard rumors that he was finishing his monumental treatise on cannibalism, I paid him a visit at Arizona State University, in Tempe, just outside Phoenix. The door to his office led me into a dim dog-leg space created by the backs of shelves and filing cabinets that stood in the middle of the room. At one end of a scarred oak table lay an untidy heap of plastic bags containing broken human bones, including skulls. Dental picks, a tiny magnifying glass, and other instruments rested nearby. The far wall was covered with photographs of skulls, of a grimacing mummy head, and of Hopi medicine men, among other things. A crude photocopy of an old sideshow bill was taped on another wall, near a computer terminal. The bill read, "The Head of the Renowned Bandit Joaquin! Will Be Exhibited for One Day Only, April 19, 1853. Plus the Hand of the Notorious Robber and Murderer Three Fingered Jack."

Turner sat behind an old desk next to a window, which overlooked a pleasant courtyard with two palm trees. He is sixty-four, and has sandy hair and bleached, watery brown eyes behind large glasses. His skin is rough and leathery from years of outdoor work in the sun.

Turner was born in Columbia, Missouri, and grew up in Southern California. "I was baptized without being consulted," he told me. "Presbyterian. I hated it so much it turned me quickly to Darwin." He had been a premedical student at the University of Arizona, but he switched to anthropology. Best known for work he has done on dental morphology—the shapes of human teeth—he has spent most of his academic life attempting to trace the various waves of human migration from Asia to America by looking at teeth.

I spent three days with Turner. He is a mercurial man, who can be by turns gracious, witty, charming, aggressive, and vituperative. It is perhaps a good thing that he has surrounded himself with the dead, because in dealing with the living he is legendarily difficult. "I have no friends, but I have no scars," he told me. He seems to relish being on the unpopular side of an academic fight. According to his daughter, Korri, "He loves it, being branded politically incorrect."

After chatting with Turner for a while, I asked him about the heap of bones on his table. He said that they were from a site called Sambrito Village, on the San Juan River, in New Mexico. It had been excavated thirty-five years before, when the area was to be drowned by Navajo

Reservoir. The excavating archaeologist had concluded that cannibalism had occurred there, but his finding was ignored. Turner was now reexamining the bones for his book.

"All the makings of cannibalism are here," he said enthusiastically, pointing to the charnel heap. He lifted a plastic bag holding a piece of skull, and slid the piece into his hand, cradling it gently. "This is a good one to illustrate the roasting of the head. A lot of the heads have this burning pattern on the back." He indicated a patch on the skull where the bone was crumbling and flaking off. He handed it to me, and I took it gingerly. "Clearly," he continued, "they were decapitating the heads and putting them in the fire face up."

"Why?" I asked. "To cook the brain?"

"It would have cooked the brain, yes," he said, rather dryly.

"What happens when a brain is cooked?"

"Thinking stops. Except among some of my students."

Turner pointed to the broken edge of the skull, which showed several places where sharp blows had opened the brainpan: small pieces of broken skull were still adhering to the edges. "These are perimortem breaks. This cannot happen except in fresh bone," he said. "Perimortem" refers to events at the time of death. Most of the bones, he said, showed numerous perimortem breaks; the crushing, splintering, and breaking of the bones had thus occurred just before, at, or just after death.

"How old was this person, and of what sex?" I asked.

Turner took the skull and flipped it over. "No signs of sutural closing. I'd put it at eighteen to twenty years. Sex is female. There are very light brow ridges, orbit is sharp, mastoid is relatively small, bone is light. I couldn't rule out a very light male."

He rummaged through the pile and showed me other bones. Some had cuts and marks of sawing near the joints, caused by dismemberment with stone tools. He pointed out similar cuts where the muscles had been attached to the bone—evidence that meat had been stripped off. He showed me percussion marks from stone choppers used to break open bones for marrow and to hack through the skulls.

Turner was the first person ever to quantify what a set of

cannibalized human remains looks like. Eventually, in the course of his work, he identified five characteristics that he felt had to be present in a bone assemblage before one could claim that the individual had been cooked and eaten:

1. Bones had to be broken open as if to get at the marrow.
2. Bones had to have cutting and sawing marks on them, made by tools, in a way that suggested dismemberment and butchering.
3. Some bones had to have "anvil abrasions." These are faint parallel scratches, which Turner noticed most often on skulls, caused when the head (or another bone) is placed on a stone that serves as an anvil, and another stone is brought down hard on it to break it open. When the blow occurs, a certain amount of slippage takes place, causing the distinctive abrasions.
4. Some bone fragments had to be burned; heads, in particular, had to show patterns of burning on the back or top, indicating that the brain was cooked.
5. Most of the vertebrae and spongy bone had to be missing. Vertebrae and spongy bone are soft and full of marrow. They can be crushed whole either to make bone cakes (something the Anasazi did with other mammal bones) or to extract grease through boiling. (Fresh bone is full of grease.)

While Turner was involved in his multi-decade project, Tim D. White, a well-known paleoanthropologist, made another discovery. In examining a six-hundred-thousand-year-old fossilized skull from Africa, he noticed some peculiar scratches. They looked as if someone had scraped and carved the flesh from the skull with a stone tool. He wondered whether the skull was evidence of cannibalism deep in the human fossil record.

To learn more about what cannibalism does to bones, White

turned his attention to the American Southwest. In 1973, in Mancos Canyon, Colorado, an archaeological team had found the broken and burned remains of approximately thirty people scattered on the floors of a small ruined pueblo. White borrowed the bones in the summer of 1985 and studied them intensively for the next five years. He found all five of the indications of cannibalism that Turner had identified. But he also noticed another peculiar trait: a faint polishing and beveling on many of the broken tips of bones. White wondered if this polishing might have been caused by the bones' being boiled and stirred in a rough ceramic pot, to render their fat. To test the idea, White and his team performed an experiment. They broke up several mule-deer bones and put the pieces in a replica of an Anasazi corrugated clay cooking pot, partly filled with water, and then heated the mixture on a Coleman stove for three hours, stirring it occasionally with a wooden stick. The fat from the bones rose to the surface and coagulated around the waterline, forming a ring of grease about half an inch thick. They decanted the contents, and White took a bone piece and scraped off the ring of fat around the inside of the pot.

Under magnification, the deer bones showed the same microscopic polishing that White had observed on the Mancos bones. Furthermore, the bone used to scrape out the ring of fat showed a pattern of scratches that exactly matched that of one Mancos bone. White called this "pot polish."

On learning of White's discovery, Turner took a second look at many of his cannibalized assemblages. He found pot polish on most of them, including the Polacca Wash bones, and he added pot polish to his list. (For the most part, Turner and White communicate with each other via scholarly journals; they have no personal relationship.)

I asked Turner if there were any examples of pot polish among the bones in his office. He searched around and pulled out a tiny fragment. He went to his desk for a handheld magnifier and examined the fragment in the brilliant Arizona light. "This is polishing," he said, with a grin, holding it up like a jewel. "I wanted you to see this. It's from Burnt Mesa, in New Mexico. Alan Brew is the excavator. The interesting thing is that the polishing occurs only on the ends of the fragments, not on the mid-portion. The physics of a pot prohibit it. We also don't get the polishing on the large pieces that won't fit into the pot."

He handed me the bone and the magnifier. As I examined it in the light, I could see a bright polished line along one fractured edge of bone.

"Just a delightful break," Turner said.

"Did you experiment with deer bones, like Tim White?" I asked, handing the bone back.

"We can't get deer in the grocery store," he said. "We used beef and chicken."

TURNER'S WORK ON Anasazi cannibalism took place during a much larger debate on cannibalism worldwide. In 1979, William Arens, a professor of anthropology at the State University of New York at Stony Brook, published an influential book entitled *The Man-Eating Myth*. The book questioned the very existence of cannibalism in human societies, and it was widely reviewed. (The *New Yorker* called it "a model of disciplined and fair argument.") Arens argued that there were no reliable, firsthand accounts of cannibalism anywhere in the historical or ethnographic record. He showed that published reports of cannibalism were mostly hearsay, from unreliable witnesses, who talked about something they had not seen personally. Despite a diligent search, he said, he had been unable to find even one anthropologist, living or dead, who claimed to have witnessed cannibalism. (Apparently, not even D. Carleton Gajdusek, who in 1976 won a Nobel Prize in Medicine for identifying a cannibalistic disease called *kuru* among the Fore tribe of New Guinea, had ever seen it.) Arens documented how some anthropologists in Brazil and elsewhere had badgered and hectored their informants until they finally "admitted" that their ancestors had been cannibals. He argued that vivid accounts of cannibalism collected by the Spanish in the Caribbean and in central Mexico were mostly written by people who were attempting to justify conquest, conversion, and enslavement.

Calling one's neighbor a cannibal, Arens went on to say, was the ultimate insult. It was always members of some tribe down the river or over the mountains who were cannibals. Or it was one's bad old ancestors, before contact with "superior" European civilization. Arens took his profession to task for not demanding more rigorous evidence before making such claims. "You have to ask anthropologists why they need cannibalism," he wrote, and he went on to give an answer: that anthropologists love cannibalism because they thrive on the exotic, the weird, the strange; they want to perpetuate the idea that some people are radically different from us and thus worth studying. Arens's book leaves little doubt that anthropologists have accepted, eagerly and uncritically, many dubious accounts of cannibalism. Naturally, when the book was published many anthropologists objected to it, but a surprising number of scientists (particularly archaeologists) felt that Arens had made a valid point, which needed to be tested. *The Man-Eating Myth* bolstered Turner's critics and contributed to an atmosphere in which his assertions were regarded with suspicion.

I called up Arens to find out what he thought of Turner's work, twenty years later. Surprisingly, he turned out to be a Turner believer. Cannibalism, he said, was "a possible interpretation, even a good interpretation," of Turner's bone assemblages. He worried, however, that most people would conclude that all the Anasazi were cannibals—and, by extension, all Native Americans. "There's a whole discipline in existence looking for 'savage' behavior among the people we have colonized, conquered, and eradicated. That point almost has to be made—that the people here before us were cannibals—to justify our genocide of Native Americans."

Turner still has many articulate detractors. One of the most outspoken is Kurt Dongoske, a white man who is the archaeologist for the Hopi tribe. Dongoske doesn't take issue with Turner's analysis of how the bones were processed, but he objects to Turner's conclusion that any people were actually eating the cooked meat. There is simply no proof that the meat was consumed, he told me, nor does he believe that Turner has sufficiently considered other alternatives, such as bizarre mortuary practices. Leigh J. Kuwanwisiwma, who is the director of the Hopi Cultural Preservation Office, wonders why Turner assumed that the Polacca Wash bones, if they were cannibalized at all, represented Hopi-on-Hopi cannibalism. He points out that Navajo, Apache, and Ute all raided the Hopi, killing men and stealing women and children. He feels that it was unfair of Turner to pursue this research without the Hopi tribe's being involved. "Turner has never sat down with us," Kuwanwisiwma says. "There was an open invitation, back in 1993, to come before the Hopi people and see if he was able to explain his research to us, and he refused. He never made contact with us before or after." Turner, for his part, says that the invitation was extended for only one visit and that he couldn't come for personal reasons. He says he offered to come at another time but never received a response. As for the possibility that other tribes had carried out these acts against the Hopi, Turner said that the Awatovi story best fitted the facts.

Others have accused Turner of insensitivity in presenting such inflammatory findings. Duane Anderson, an archaeologist who is the vice-president of the highly respected School of American Research, in Santa Fe [now the School for Advanced Research], has said that Turner, in common with some other physical anthropologists, doesn't show much concern about how his work might affect living, related people. "There's a tendency when dealing with bones to treat the material as objects, rather than as subjects."

Many of Turner's critics have proposed alternative explanations. J. Andrew Darling, the executive director for the Mexico-North Research Network, in Chihuahua, wrote a paper, "Mass Inhumation and the Execution of Witches in the American Southwest," published in 1998, suggesting that the bones might be those of witches executed in a particularly grisly fashion. The utter destruction of the witch's body through dismemberment, defleshing, burning, boiling, and crushing was an attempt to efface his or her evil powers. He cites known instances of Pueblo Indian witches being killed, violently mutilated, and dismembered. Debra L. Martin, a professor of biological anthropology at Hampshire College, in Massachusetts, and an authority on Anasazi violence, also feels that Turner rejected other explanations out of hand. "I don't see why those bones couldn't have been stomped on, cut up, broken apart, and boiled ritually" without being eaten, she said.

"And why isn't there anything about cannibalism in the ethnographic record? I would like to see just one clan history, one story, one early Spanish account, that confirms this." Martin herself examined some of Turner's bone assemblages. "Christy homogenizes all these assemblages in his publications," she pointed out. "He makes them all seem alike. But they're not. What if the explanation is a lot more interesting? What if it's something grander than this? Maybe some cannibalism, some witchcraft executions, and some really unusual or interesting mortuary practices?" She made another point. Because Turner did not collaborate with Native Americans, "they are making a special effort to reclaim and bury those bones," she said. "He'll have been the only person who's looked at them, and that's too bad."

Others are even more blunt in describing Turner. "He's not nice," one colleague said. "He's a pain in the ass." Another called him "loud" and "a bully." There is a high level of apprehension at some of the museums where Turner has done his research. At the Museum of Northern Arizona, his longtime hangout, I was firmly denied permission even to enter the collection area, let alone look at the bones Turner had worked on. "We're walking a tightrope here," Noland Wiggins, a collections supervisor, said apologetically. "Because, as you may imagine, the tribes are not too happy with Christy's research." Turner doesn't shy away from responding forcefully to his critics. In a recent paper he accused one of them of "playing to the choir" and called Dongoske "self-serving." He also considered it strange that his critics and the Hopi were more exercised over cannibalism than over violence and mutilation. "How can you tolerate killing, murder, and torture and then be so horrified by cannibalism? Why is it that the Hopi can admit killing eight hundred at Awatovi as if it were nothing, but then the whole universe falls apart when they are accused of cannibalism?"

I asked him why he felt that there had been so much opposition to his ideas. "There's a simple answer," he said, with a mirthless smile. "In our culture, cannibalism is a food taboo. That's the essence of this whole problem."

Although Turner admitted that he had no direct evidence that human meat was eaten at any of the sites, he said that he based his fundamental conclusion of cannibalism on the scientific principle of Occam's razor: the simplest explanation fitting the facts is probably the right one. "There's still a chance," he said sarcastically, "that aliens are doing this." Nevertheless, his critics continue to point out that he lacks proof. In 1996, Kurt Dongoske was quoted in *National Geographic* as saying that Anasazi cannibalism would not be proved until "you actually find human remains in prehistoric human excrement."

IN THE EARLY 1990s, a firm called Soil Systems won a contract to excavate a group of archaeological sites at the base of Sleeping Ute Mountain, in Colorado, on the Ute Mountain Ute Indian Reservation. The Ute planned to irrigate and farm 7,600 acres of land, and the law required them to excavate any archaeological sites that would be disturbed. The project director at Soil Systems was a young man named Brian Billman, who is now an assistant professor at the University of North Carolina at Chapel Hill.

He and his team began work in 1992, and at one unremarkable site along Cowboy Wash, called 5MT 10010, he and two colleagues, Patricia Lambert and Banks Leonard, made a grotesque discovery. The results have not yet been published, but Billman was willing to talk to me about them—up to a point. [The results were published in *American Antiquity* in January 2000.] We spoke by telephone the day before he was to go off to Peru to do fieldwork. Billman spoke slowly and carefully, weighing every word, and this is the story he told:

When the team began excavating, they uncovered what seemed at first a typical Anasazi site—some rooms, a trash mound, and, lined up in a row, three kivas. As the team dug out the first kiva, they found a pile of chopped-up, boiled, and burned human bones at the base of a vent shaft leading up and out of the kiva. It looked as though the bones had been chopped up and cooked outside, on the surface, and then dumped down the shaft. There were cut marks on the bones made by stone tools, and the long bones had been systematically broken up for marrow extraction.

In the second kiva, they found the remains of five individuals. In this case, it appeared that the bones had been processed inside the kiva itself. "Instead of boiling," Billman recalled, "it looked more like roasting going on." Here cut marks at muscle attachments suggested that the bones had been defleshed, and again they had been split open for marrow. The skulls of at least two of the individuals had been placed upside down on the fire, roasted, and broken open, and the cooked brains presumably scooped out. In that same kiva, the team found a stone tool kit such as was typically used in butchering a midsize mammal. The kit contained an axe, hammerstones, and two large flakes with razor-thin cutting edges. Billman submitted the tool kit to a lab, and the two flakes tested positive for human blood.

The third kiva contained only two small pieces of bone, which had apparently been washed down from the surface. In the dead ashes of the central hearth, however, the team made an "extremely unusual" find. It was a nondescript lump of some material, which was field-classified as a "macrobotanical remain"—a piece of an unidentified plant. A worker put it in a bag, and when the team had a chance to examine it more closely, back in the laboratory, they realized that it was a desiccated human turd, or coprolite. "After the fire had gone cold," Billman said, "someone had squatted over this hearth and defecated into it."

Billman sent the coprolite off to a lab at the University of Nebraska for analysis. The first oddity the lab noted was that it contained no plant remains; other tests indicated that the coprolite had formed from digested meat. From a pollen analysis, the lab could tell that the coprolite had been deposited in the late spring or early summer, at the same time of year that the site was abandoned.

In three nearby ruined sites, another group of excavators also found chopped-up, boiled, and burned bones scattered about. The four sites, which seemed to constitute a small community, contained a total of twenty-eight butchered individuals. Mysteriously, all four sites were filled with valuable, portable items, such as baskets, a rabbit blanket, pots, and ground-stone tools. Little, if anything, seemed to have been taken.

"This site has a frozen instant of time in it," Billman said. "You could almost read it." What he read was: The year was approximately 1150. Times were hard. The area was in the grip of a severe drought. Pollen samples showed that a crop failure had probably occurred the previous year. One late-spring day, the community was attacked. The people were killed, cooked, and eaten. Then, in an ultimate act of contempt, one of the killers defecated in a hearth, the symbolic center of the family and the household. Instead of looting the site, the invaders left it and its many valuables for all to see.

"When I excavated it," Billman told me, "I got the sense that it may have been taboo. We are proposing that this may have been a political strategy. One or several communities in this area may have used raiding and cannibalism to drive off people from a village and prevent other people from settling there. If you raided a village, consumed some of the residents, and left the remains there for everyone to see, you would gain the reputation of being a community to stay away from."

Billman, Lambert, and Leonard presented their findings at the 1997 Society for American Archaeology meetings, in Nashville, and they were subsequently reported by Catherine Dold in *Discover*. At the end of the Nashville symposium, a man came up to Billman, introduced himself as Richard Marlar, and said, "I'm a biochemist, and I think I can tell you if there is human tissue in that coprolite"—in other words, he could determine directly whether or not cannibalism had occurred. Billman sent some samples of the coprolite off to Marlar for analysis, along with some pieces of a ceramic Anasazi cooking pot found at the site.

I recently called up Richard Marlar, who is an associate professor of pathology at the University of Colorado Health Sciences Center, in Denver. "I heard his talk," Marlar said. "I said to myself, 'I can figure that out. We can answer that question.'" The basic problem, he realized, was that he needed a way to identify human tissue that had passed through the digestive system of another human being. He had to make sure he was not picking up traces of human blood in the intestinal tract (from internal bleeding) or cells naturally shed from the lining of the intestine. He finally decided to test the coprolite for the presence of human myoglobin, a protein that is found only in skeletal and heart muscle, and could not get into the intestinal tract except through eating. (As a control, Marlar tested many stool samples from patients in his hospital, to verify that none had traces of myoglobin in them.)

Marlar set up an immunological assay of the kind that is normally used in clinical medicine to determine whether someone has a disease. So far, he explained to me, he had performed seven tests, each in triplicate, using twenty-one samples of the coprolite. He also ran six tests on the ceramic pottery to see if it had traces of human protein from cooking. All the results, he told me, were the same.

"And what were those results?" I asked.

He declined to answer. The Ute tribe had asked all the excavators to keep the results confidential until the paper could be published.

In the small world of Southwestern archaeology, very little can be kept secret, and I soon began to hear rumors about the results. I tracked them down and established that the tests had been positive. All of Marlar's assays, I learned, had shown the presence of human myoglobin protein in the coprolite and on the interior walls of the cooking pot.

FOR THIRTY YEARS, Turner had been documenting cannibal sites, but for a long time he had not tackled the question "Why?" It is this question that he takes up in the last, and what is certain to be the most controversial, chapter of *Man Corn*. He advances a theory of who the cannibals were, where they came from, and what role the eating of "men, women, and children alike" may have played in Anasazi society. For this was not, he says, starvation cannibalism, such as befell the Donner party. Starvation cannibalism did not explain the extreme mutilation of the bodies before they were consumed, or the huge charnel deposits, consisting of as many as thirty-five people (that's almost a ton of edible human meat), or the bones discarded as trash. Furthermore, there was no evidence of starvation cannibalism (or any other kind of cannibalism) among the Anasazi's immediate neighbors, the

Hohokam and the Mogollon, who lived in equally harsh environments and endured the same droughts.

A colleague of Turner's, David Wilcox, who is a curator at the Museum of Northern Arizona, had prepared a map showing the distribution of Chaco Great Houses and roads. Using Wilcox's map, Turner was able to chart charnel deposits in time and space.

"When we found that Dave's Chaco maps coincided with my cannibalized assemblages," Turner recalls, "that's when it came together." Turner decided that the civilization centered in Chaco Canyon was probably the locus of Anasazi cannibalism.

The maps, Turner says, showed that the charnel deposits were often situated near Chaco Great Houses and that most of them dated from the Chaco period. The eating of human flesh seems to have begun as the Chaco civilization began, around 900; peaked at the time of the Chaco collapse and abandonment, around 1150; and then all but disappeared (Polacca Wash being a notable exception).

Turner theorized that cannibalism might have been used by a powerful elite at Chaco Canyon as a form of social control. "It was order by terrorism," he said to me. "Big-stick order." In *Man Corn* he writes:

> Terrorizing, mutilating, and murdering might be evolutionarily useful behaviors when directed against unrelated competitors. And what better way to amplify opponents' fear than to reduce victims to the subhuman level of cooked meat, especially when they include infants and children from whom no power or prestige could be derived but whose consumption would surely further terrorize, demean, and insult their helpless parents or community.... The benefits would be threefold: community control, control of reproductive behavior (that is, dominating access to women), and food. From the standpoint of sociobiology, then, cannibalism could well represent useful behavior done by well-adjusted, normal adults acting out their ultimate, evolutionarily channeled behavior. On the other hand, one can easily look upon violence and cannibalism as socially pathological.

The second question Turner asked was "Who were these cannibals and where did they come from?" He looked around for a source. The Anasazi's immediate neighbors showed no evidence of being cannibalistic. "I couldn't find cannibalism in California or on the Great Plains, either," Turner said. "Where is it? In Mexico."

Turner directed his attention to central Mexico, to the empire of the Toltecs—the precursors of the Aztecs—which lasted from about 800 to 1100 A.D. Central Mexico, he writes, developed a "very powerful, dehumanizing sociopolitical and ideological complex," centered on human sacrifice and cannibalism used as a form of social control. Furthermore, cannibalism spread from central Mexico "into the jungle world of the Mayas and the desert world of Chichimeca" in northern Mexico. Turner concludes, "It takes nearly blind faith in the effectiveness of geographical distance... to believe that this complex and its adherents failed to reach the American Southwest."

During the Toltec period, Turner hypothesizes, a heavily armed group of "thugs," "tinkers," or perhaps even "Manson party types" (as he put it to me in various conversations) headed north, to the region we refer to as the American Southwest. "They entered the San Juan Basin around A.D. 900," he surmises in *Man Corn*, and "found a suspicious but pliant population whom they terrorized into reproducing the theocratic lifestyle they had previously known in Mesoamerica."

In other words, the flowering of Chaco society that we have so long admired—in engineering, astronomy, architecture, art, and culture—was the product of a small, heavily armed gang from Mexico, who marched into the Southwest to conquer and brutalize.

Archaeologists have long known that there was a strong Mesoamerican influence on the Anasazi. There was extensive prehistoric trade between Mexico and the Southwest. Turquoise from Santa Fe has been found throughout Mesoamerica, and tropical parrots and macaws brought up live from Mexico have been found in Chaco graves. Indeed, corn, pottery, and cotton originally came into the Southwest from Mexico. There is good evidence, Turner writes in *Man Corn*, that Mexicans did in fact make the journey northward. He notes that a skull found in Chaco Canyon had intentionally chipped teeth—a decorative trait thought to be restricted almost entirely to central Mexico. He also details many parallels between Hopi and Toltec mythology.

A number of Pueblo Indian myths seem to support Turner's cannibalism theories. For example, a Pueblo legend collected by the anthropologist John Gunn and published in 1916 describes a drought and famine in the past which reduced the people "to such an extremity that they killed and ate their children or weaker members of the tribe."

The Navajo tell many stories about Chaco Canyon that paint a very different picture from the popular Anglo view—stories that may also have been taken from the Pueblo Indians. While doing research for a book on the Navajo creation story, I was told a number of these stories. Chaco, some older Navajo say, was a place of hideous evil. The Chaco people abused sacred ceremonies, practiced witchcraft and cannibalism, and made a dreaded substance called corpse powder by cooking and grinding up the flesh and bones of the dead. Their evil threw the world out of balance, and they were destroyed in a great earthquake and fire.

CANNIBALISM SEEMS TO have peaked in the Southwest at the time of the Chaco collapse because the system of terror, Turner theorizes, could not be sustained. Terror begat social chaos. "The evidence is that cannibalism—and this chaos that ensued—started in the north and it rippled southward and it wiped the Southwest out," Turner told me. In other words, cannibalism and social terror may have been a factor—perhaps the *missing* factor—in the Chaco collapse. Turner doesn't reject the standard explanations of the Chaco implosion: he hypothesizes that social pathology and cannibalism, combined with one or more of the standard theories (drought, erosion, disease, famine), sparked chaos, violence, and the "near-extinction of the entire prehistoric Southwest population." The subsequent retreat of the Anasazi into inaccessible cliff dwellings and mesas now makes sense. The long-sought elusive enemy of the Anasazi was, in fact, themselves.

Turner gave me a paper he had just written and was planning to deliver at a conference. Entitled "The Darker Side of Humanity," the paper extends some of the ideas in the last chapter of *Man Corn*. Turner writes, "I can easily imagine the cancerous random fractals of social chaos branching all over the Southwest, starting in the north with the collapse of Chaco and like a wildfire erupting here and there in hot spots of human violence.... Think of the hundreds of thousands of socially pathological killings and mutilations committed in central Africa these last few years. Think about Pol Pot."

Cannibalism was not "normal" behavior among the Anasazi, he argues, even if it was widespread. It was the product of a few socially pathological individuals who whipped up the emotions of their followers, like the chief of Awatovi who plotted the grim extinction of his own village. Turner compares such men as the Awatovi chief to Adolf Hitler, Genghis Khan, and Joseph Stalin.

This argument leads Turner into even stranger territory. In his paper he calls on archaeologists to give up the time-honored "concept of culture." The problem with archaeology is that it is a science of generalization. The archaeologist digs a site and then extrapolates the findings into a description of a culture. The orientation of the archaeologist is always toward matters like "What was usual and customary in this culture?" and "What was the norm?" There is no provision for abnormality, for the charismatic or sociopathic individual—the deranged Great Man. "In my thirty-five years of teaching I have never heard of a graduate student specializing in archaeology who had taken a course or a seminar in abnormal psychology," Turner writes. "Why should they? ... The very idea of abnormal behavior is alien to Southwest archaeological thinking." He suggests replacing the paradigm of culture with a "Darwinian paradigm of evolutionary psychology" that "emphasizes identification of individuals and seeks to understand their actions wherever possible." Only through this paradigm shift, Turner asserts, will archaeologists be able to understand the darker side of human nature in the archaeological record.

ON MY LAST day with Turner, he decided to visit a cannibal locale in Monument Valley, straddling the border of Utah and Arizona.

We left the cool ponderosas of Flagstaff on a June morning. By three o'clock, we had arrived at the escarpment that looks down on Monument Valley—surely one of the most dramatic landscapes on earth. We descended into the valley on a rutted dirt road, our two cars kicking up corkscrews of red dust, and after a few miles the Three Sisters came into view on the right, three spires of rock. According to a photograph that Turner had of the site—his only clue to where it was—it lay less than a mile from this valley landmark. As we bounced along the valley floor, the spires began to move into the alignment seen in the photograph.

Turner lurched off the road to follow a track in the bottom of a dry wash. We skirted the base of a large mesa, stopped, backtracked, and stopped again. Turner finally got out, squinting in the brilliant sunlight and clutching the photograph. "This is it," he said. "This is it, exactly."

We scrambled up the sandy rise above the wash. The site lay about twenty feet above the valley floor, on the talus slope of Thunderbird Mesa. It was a small patch of sand sheltered among giant plates of stone that had spalled from the cliff behind—a sheer wall of red sandstone four hundred feet high and streaked with glossy desert varnish. It was a breathtaking spot, commanding a sweeping view of Tse Biyi Flats, Rain God Mesa, the Three Sisters, Spearhead Mesa, and dozens of other buttes and mesas layered one against another, receding into vast distances. The afternoon sun was invading the valley, sculpting and modeling the buttes in crisp yellow light.

The site itself was covered with windblown sand and clumps of Indian rice grass and snakeweed. In the center lay a large, exquisite piece of a painted Anasazi pot, white with a black geometric design. Near the pot, the edge of a slab-lined hearth stuck up from the sand. The smashed, chopped, and burned bones of seven people had been found piled in this hearth. They were the remains of an old man and an old woman, a younger man, two teenage girls, a third adolescent, of undetermined sex, and an infant. Turner believed that they had been ambushed, killed, mutilated, dismembered, and cooked right there, for the hearth seemed to have been custom-built for that purpose.

After it had been used, the cracked and burned bones were left there in the fire pit and the site was abandoned.

Turner poked around the site, scowling and squinting, with two cameras swinging from his neck. "Dogoszhi black-on-white," he said, glancing at the potsherd and referring to a common Anasazi pottery type. He took a careful series of photographs of the site and its surroundings.

"Do you think that potsherd was left at the time of the massacre?"

"Yes," he replied.

"Why here?"

"That's a bit of a mystery. It's not *at* anything. But I wouldn't be surprised if there was a Great House near here, and somebody got waylaid." He pointed to a wedge of green growth nearby. "There must be an intermittent spring there. That would be part of the story—perhaps this was a hunting camp for deer or antelope. Perhaps it was in wintertime. This is a nice place in the winter."

As we were tramping around the site, an old Navajo man came by in a pickup truck, which had two dazed, dust-covered tourists in the back. He was wearing a straw cowboy hat and was missing his front teeth. He stopped the truck.

"Any Anasazi ruins around here?" Turner called out.

"Over there," the man said, his hand waving obscurely across thousands of desolate acres. He seemed reluctant to talk more about the Anasazi, and drove on.

"That fellow was vague about ruins," Turner said to me. "But there must be some nearby. This is a *chindi* place"—a place of ghosts. He continued to move restlessly around the site—a skinny man with a potbelly and sticklike arms and legs—staring into every recess. Only the occasional click of his camera broke the stillness. I remembered my first interview with Turner, when I had asked him why he was investigating cannibalism. He had replied breezily, "I think it's interesting. It's fun. Here's an unsolved problem." As I looked at his face, I could see that he was indeed having a marvelous time.

Turner moved back to the car. I remained at the spot and looked around, trying to arrive at an understanding of what had happened here. The age and sex of the remains suggested that they might have been an extended family—two parents, three teenage children, a son-in-law, and a little grandchild, perhaps. I thought of my own family. The light deepened. A grasshopper began scratching among the dry stones, and a faint breeze brought with it the scent of sun-warmed sand.

UPDATE

Since the article was published, the term "Anasazi" has fallen out of favor because it is a Navajo word with negative connotations. It has been replaced by the more accurate term Ancestral Puebloans.

Christy Turner died in 2013. His work, while still considered controversial, has been generally accepted by the profession. While many theories have been offered as to what caused the outbreak of violence and cannibalism in the Southwest during that time, no clear answer has emerged.

THE LOST TOMB

ON FEBRUARY 2, 1995, at ten in the morning, the archaeologist Kent R. Weeks found himself a hundred feet inside a mountain in Egypt's Valley of the Kings, on his belly in the dust of a tomb. He was crawling toward a long-buried doorway that no one had entered for at least thirty-one hundred years. There were two people with him, a graduate student and an Egyptian workman; among them they had one flashlight.

To get through the doorway, Weeks had to remove his hard hat and force his large frame under the lintel with his toes and fingers. He expected to enter a small, plain room marking the end of the tomb. Instead, he found himself in a vast corridor, half full of debris, with doorways lining either side and marching off into the darkness. "When I looked around with the flashlight," Weeks recalled later, "we realized that the corridor was tremendous. I didn't know *what* to think." The air was dead, with a temperature in excess of a hundred degrees and a humidity of one hundred percent. Weeks, whose glasses had immediately steamed up, was finding it hard to breathe. With every movement, clouds of powder arose, and turned into mud on the skin.

The three people explored the corridor, stooping, and sometimes crawling over piles of rock that had fallen from the ceiling. Weeks counted twenty doorways lining the hundred-foot hallway, some opening into whole suites of rooms with vaulted ceilings carved out of

---

Originally published in the *New Yorker* in 1996.

the solid rock of the mountain. At the corridor's end, the feeble flashlight beam revealed a statue of Osiris, the god of resurrection: he was wearing a crown and holding crossed flails and sceptres; his body was bound like that of a mummy. In front of Osiris, the corridor came to a T, branching into two transverse passageways, each of them eighty feet long and ending in what looked like a descending staircase blocked with debris. Weeks counted thirty-two additional rooms off those two corridors.

The tomb was of an entirely new type, never seen by archaeologists before. "The architecture didn't fit any known pattern," Weeks told me. "And it was so *big*. I just couldn't make sense of it." The largest pharaonic tombs in the Valley contain ten or fifteen rooms at most. This one had at least sixty-seven—the total making it not only the biggest tomb in the Valley but possibly the biggest in all Egypt. Most tombs in the Valley of the Kings follow a standard architectural plan—a series of consecutive chambers and corridors like a string of boxcars shot at an angle into the bedrock and ending with the burial vault. This tomb, with its T shape, had a warren of side chambers, suites, and descending passageways. Weeks knew from earlier excavations that the tomb was the resting place for at least four sons of Ramesses II, the pharaoh also known as Ramesses the Great—and, traditionally, as simply Pharaoh in the Book of Exodus. Because of the tomb's size and complexity, Weeks had to consider the possibility that it was a catacomb for as many as fifty of Ramesses' fifty-two sons—the first example of a royal family mausoleum in ancient Egypt.

Weeks had discovered the tomb's entrance eight years earlier, after the Egyptian government announced plans to widen the entrance to the Valley to create a bus turnaround at the end of an asphalt road. From reading old maps and reports, he had recalled that the entrance to a lost tomb lay in the area that was to be paved over. Napoleon's expedition to Egypt had noted a tomb there, and a rather feckless Englishman named James Burton had crawled partway inside it in 1825. A few years later, the archaeologist Sir John Gardner Wilkinson had given it the designation KV5, for Kings' Valley Tomb No. 5, when he numbered eighteen tombs there. Howard Carter—the archaeologist who had discovered King Tutankhamun's tomb in 1922, two hundred feet further on—dug two feet in, decided that KV5's entrance looked unimportant, and used it as a dumping ground for debris from his other excavations, thus burying it under ten feet of stone and dirt. The location of the tomb's entrance was quickly forgotten.

It took about ten days of channeling through Carter's heaps of debris for Weeks and his men to find the ancient doorway of KV5, and it proved to be directly across the path from the tomb of Ramesses the Great. The entrance lay at the edge of the asphalt road, about ten feet below grade and behind the rickety booths of T-shirt vendors and fake-scarab-beetle sellers.

Plans for the bus turnaround were canceled, and, over a period of seven years, Weeks and his workmen cleared half of the first two chambers and briefly explored a third one. The tomb was packed from floor to ceiling with dirt and rocks that had been washed in by flash floods. He uncovered finely carved reliefs on the walls, which showed Ramesses presenting various sons to the gods, with their names and titles recorded in hieroglyphics. When he reached floor level, he found thousands of objects: pieces of faience jewelry, fragments of furniture, a wooden fist from a coffin, human and animal bones, mummified body parts, chunks of sarcophagi, and fragments of the canopic jars used to hold all the mummified organs of the deceased—all detritus left by ancient tomb robbers.

The third chamber was anything but modest. It was about sixty feet square, one of the largest rooms in the Valley, and was supported by sixteen massive stone pillars arranged in four rows. Debris filled the room to within about two feet of the ceiling, allowing just enough space for Weeks to wriggle around. At the back of the chamber, in the axis of the tomb, Weeks noticed an almost buried doorway. Still believing that the tomb was like the others in the Valley, he assumed that the doorway merely led to a small, dead-end annex, so he didn't bother with it for several years—not until February 1995, when he decided to have a look.

Immediately after the discovery, Weeks went back to a four-dollar-a-night pension he shared with his wife, Susan, in the mud village of Gezira Bairat, showered off the tomb dust, and took a motorboat across the Nile to the small city of Luxor. He faxed a short message to Cairo, three hundred miles downriver. It was directed to his major financial supporter, Bruce Ludwig, who was attending a board meeting in the American University in Cairo, where Weeks is a professor. It read, simply, "Have made wonderful discovery in Valley of the Kings. Await your arrival." Ludwig instantly recognized the significance of the fax and the inside joke it represented: it was a close paraphrase of the telegram that Howard Carter had sent to the Earl of Carnarvon, his financial supporter, when he discovered Tutankhamun's tomb. Ludwig booked a flight to Luxor.

"That night, the enormousness of the discovery began to sink in," Weeks recalled. At about two o'clock in the morning, he turned to his wife, and said, "Susan, I think our lives have changed forever."

The discovery was announced jointly by Egypt's Supreme Council of Antiquities, which oversees all archaeological work in the country, and the American University in Cairo, under whose aegis Weeks was working. It became the biggest archaeological story of the decade, making the front page of the *Times* and the cover of *Time*. Television reporters descended on the site. Weeks had to shut down the tomb to make the talk show circuit. The London newspapers had a field day: the *Daily Mail* headlined its story "PHARAOH'S 50 SONS IN MUMMY OF ALL TOMBS," and one tabloid informed its readers that texts in the tomb gave a date for the Second Coming and the end of the world, and also revealed cures for AIDS and cancer.

The media also wondered whether the tomb would prove that Ramesses II was indeed the pharaoh referred to in Exodus. The speculation centered on Amun-her-khepeshef, Ramesses' firstborn son, whose name is prominent on KV5's walls. According to the Bible, in order to force Egypt to free the Hebrews from bondage the Lord visited a number of disasters on the land, including the killing of all firstborn Egyptians from the pharaoh's son on down. Some scholars believe that if Amun-her-khepeshef's remains are found it may be possible to show at what age and how he died.

Book publishers and Hollywood producers showed a great interest in Weeks's story. He didn't respond at first, dismissing inquiries with a wave of the hand. "It's all *kalam fadi*," he said, using the Arabic phrase for empty talk. Eventually, however, so many offers poured in that he engaged an agent at William Morris to handle them; a book proposal will be submitted to publishers later this month. [The book, *The Lost Tomb*, was published by William Morrow in 1998.]

In the fall, Weeks and his crew decided to impose a partial media blackout on the excavation site—the only way they could get any work done, they felt—but they agreed to let me accompany them near the end of the digging season. Just before I arrived, in mid-November, two mysterious descending corridors, with dozens of new chambers, unexpectedly came to light, and I had the good fortune to be the only journalist to see them.

THE VALLEY OF the Kings was the burial ground for the pharaohs of the New Kingdom, the last glorious period of Egyptian history. It began around 1550 B.C., when the Egyptians expelled the foreign Hyksos rulers from Lower Egypt and re-established a vast empire, stretching across the Middle East to Syria. It lasted half a millennium. Sixty years before Ramesses, the pharaoh Akhenaten overthrew much of the Egyptian religion and decreed that thenceforth Egyptians should worship only one god—Light, whose visible symbol was Aten, the disk of the sun. Akhenaten's revolution came to a halt at his death. Ramesses represented the culmination of the return to tradition. He was an exceedingly conservative man, who saw himself as the guardian of the ancient customs, and he was particularly zealous in erasing the heretic pharaoh's name from his temples and stelae, a task begun by his father, Seti I. Because Ramesses disliked innovation, his monuments were notable not for their architectural brilliance but for their monstrous size. The New Kingdom began a slow decline following his rule, and finally sputtered to an end with Ramesses XI, the last pharaoh buried in the Valley of the Kings.

The discovery of KV5 will eventually open for us a marvelous window on this period. We know almost nothing about the offspring of the New Kingdom pharaohs or what roles they played. After each eldest prince ascended the throne, the younger sons disappeared so abruptly from the record that it was once thought they were routinely executed. The burial chambers' hieroglyphics, if they still survive, may give us an invaluable account of each son's life and accomplishments. There is a remote possibility—it was suggested to me by the secretary-general of the Supreme Council of Antiquities, Professor Abdel-Halim Nur el-Din, who is an authority on women in ancient Egypt—that Ramesses' daughters might be buried in KV5 as well. (Weeks thinks the possibility is highly unlikely.) Before Weeks is done, he will probably find sarcophagi, pieces of funerary offerings, identifiable pieces of mummies, and many items with hieroglyphics on them. The tomb will add a new chapter to our understanding of Egyptian funerary traditions. And there is always a possibility of finding an intact chamber packed with treasure.

Ramesses the Great's reign lasted an unprecedented sixty-seven years, from 1279 to 1213 B.C. He covered the Nile Valley from Nubia to the delta with magnificent temples, statuary, and stelae, which are some of the grandest monuments the world has ever seen. Among his projects were the enormous forecourt at Luxor Temple; the Ramesseum; the cliffside temples of Abu Simbel; the great Hall of Columns at Karnak; and the city of Pi-Ramesse. The two "vast and trunkless legs of stone" with a "shattered visage" in Shelley's poem "Ozymandias" were those of Ramesses—fragments of the largest statue in pharaonic history. Ramesses outlived twelve of his heirs, dying in his early nineties. The thirteenth crown prince, Merneptah, became pharaoh only in his sixties.

By the time Ramesses ascended the throne, at twenty-five, he had fathered perhaps ten sons and as many daughters. His father had started him out with a harem while he was still a teenager, and he had two principal wives, Nefertari and Istnofret. He later added several Hittite princesses to his harem, and probably his sister and two daughters. It is still debated whether the incestuous marriages of the pharaohs were merely ceremonial or actually consummated. If identifiable remains of Ramesses' sons are found in KV5, it is conceivable that DNA testing might resolve this vexing question.

In most pharaonic monuments we find little about wives and children, but Ramesses showed an unusual affection for his family, extolling the accomplishments of his sons and listing their names on numerous temple walls. All over Egypt, he commissioned statues of Nefertari (not to be confused with the more famous Nefertiti, who was Akhenaten's wife), "for whose sake the very sun does shine." When she died, in Year 24 of his reign, Ramesses interred her in the most beautiful tomb yet discovered in the Valley of the Queens, just south of the Valley of the Kings. The tomb survived intact, and its incised and painted walls are nearly as fresh as the day they were fashioned. The rendering of Nefertari's face and figure perhaps speaks most eloquently of Ramesses' love for her. She is shown making her afterlife journey dressed in a diaphanous linen gown, with her slender figure emerging beneath the gossamer fabric. Her face was painted using the technique of chiaroscuro—perhaps the first known example in the history of art of a human face being treated as a three-dimensional volume. The Getty Conservation Institute recently spent millions restoring the tomb. The Getty recommended that access to the tomb be restricted, in order to preserve it, but the Egyptian government has opened it to tourists, at thirty-five dollars a head.

The design of royal tombs was so fixed by tradition that they had no architect, at least as we use that term today. The tombs were laid out and chiseled from ceiling to floor, resulting in ceiling dimensions that are precise and floor dimensions that can vary considerably. All the rooms and corridors in a typical royal tomb had names, many of which we still do not fully understand: the First God's passage, Hall of Hindering, Sanctuaries in Which the Gods Repose. The burial chamber was often called the House of Gold. Some tombs had a Hall of Truth, whose murals showed the pharaoh's heart being weighed in judgment by Osiris, with the loathsome god Ammut squatting nearby, waiting to devour it if it was found wanting. Many of the reliefs were so formulaic that they were probably taken from copybooks. Yet even within this rigid tradition breathtaking flights of creativity and artistic expression can be found.

Most of the tombs in the Valley were never finished: they took decades to cut, and the plans usually called for something more elaborate than the pharaoh could achieve during his rule. As a result, the burial of the pharaoh was often a panicky, ad-hoc affair, with various rooms in the tomb being adapted for other purposes, and decorations and texts painted in haste or omitted completely. (Some of the most beautiful inscriptions were those painted swiftly; they have a spontaneity and freshness of line rivaling Japanese calligraphy.)

From the time of Ramesses II on, the tombs were not hidden: their great doorways, which were made of wood, could be opened. It is likely that the front rooms of many tombs were regularly visited by priests to make offerings. This may have been particularly true of KV5, where the many side chambers perhaps served a purpose. The burial chambers containing treasure, however, were always sealed.

Despite all the monuments and inscriptions that Ramesses left us, it is still difficult to bridge the gap of thirty-one hundred years and see Ramesses as a person. One thing we do know: the standard image of the pharaoh, embodied in Shelley's "frown, and wrinkled lip, and sneer of cold command," is a misconception. One of the finest works from Ramesses' reign is a statue of the young king now in the Museo Egizio, in Turin. The expression on his face is at once compassionate and otherworldly, not unlike that of a Giotto Madonna; his head is slightly bowed, as if to acknowledge his role as both leader and servant. This is not the face of a tyrant-pharaoh who press-ganged his people into building monuments to his greater glory. Rather, it is the portrait of a ruler who had his subjects' interests at heart, and this is precisely what the archaeological and historical records suggest about Ramesses. Most of the Egyptians who labored on the pharaoh's monuments did so proudly and were, by and large, well compensated. There is a lovely stela on which Ramesses boasts about how much he has given his workers, "so that they work for me with their full hearts." Dorothea Arnold, the head curator at the Egyptian Department at the Metropolitan Museum, told me, "The pharaoh was *believed* in. As to whether he was beloved, that was beside the point: he was *necessary*. He was life itself. He represented everything good. Without him there would be nothing."

Final proof of the essential humanity of the pharaonic system is that it survived for more than three thousand years. (When Ramesses ascended the throne, the pyramids at Giza were already thirteen hundred years old.) Egypt produced one of the most stable cultural and religious traditions the world has ever seen.

VERY LITTLE LIVES in the Valley of the Kings now. It is a wilderness of stone and light—a silent, roofless sepulchre. Rainfall averages a quarter inch per year, and one of the hottest natural air temperatures on earth was recorded in the surrounding mountains. And yet the Valley is a surprisingly intimate place. Most of the tombs lie within a mere forty acres, and the screen of cliffs gives the area a feeling of privacy. Dusty paths and sun-bleached, misspelled signs add a pleasant, ramshackle air.

The Valley lies on the outskirts of the ancient city of Thebes, now in ruins. In a six-mile stretch of riverbank around the city, there are as many temples, palaces, and monuments as anywhere else on earth, and the hills are so pockmarked with the yawning pits and doorways of ancient tombs that they resemble a First World War battlefield. It is dangerous to walk or ride anywhere alone. Howard Carter discovered an important tomb when the horse he was riding broke through and fell into it. Recently, a Canadian woman fell into a tomb while hiking and fractured her leg; no one could hear her screams, and she spent the days leading up to her death writing postcards. One archaeologist had to clear a tomb that contained a dead cow and twenty-one dead dogs that had gone in to eat it.

Almost all the tombs lying open have been pillaged. A papyrus now in Italy records the trial of someone who robbed KV5 itself in 1150 B.C. The robber confessed under torture to plundering the tomb of Ramesses the Great and then going "across the path" to rob the tomb of his sons. Ancient plunderers often vandalized the tombs they robbed, possibly in an attempt to destroy the magic that supposedly protected them. They smashed everything, levered open sarcophagi, ripped apart mummies to get at the jewelry hidden in the wrappings, and sometimes threw objects against the walls with such force that they left dents and smudges of pure gold.

Nobody is sure why this particular valley, three hundred miles up the Nile from the pyramids, was chosen as the final resting place of the New Kingdom pharaohs. Egyptologists theorize that the sacred pyramidal shape of el-Qurn, the mountain at the head of the Valley of the Kings, may have been one factor. Another was clearly security: the Valley is essentially a small box canyon carved out of the barren heart of a desert mountain range; it has only one entrance, through a narrow gorge, and the surrounding cliffs echo and magnify any sounds of human activity such as the tapping of a robber's pick on stone.

Contrary to popular belief, the tombs in the Valley are not marked with curses. King Tut's curse was invented by Arthur Weigall, an Egyptologist and journalist at the *Daily Mail*, who was furious that Carnarvon had given the London *Times* the exclusive on the discovery. Royal tombs did not need curses to protect them. Priests guarded the Valley night and day, and thieves knew exactly what awaited them if they were caught: no curse could compete with the fear of being impaled alive. "There are a few curses on some private tombs and in some legal documents," James Allen, an Egyptologist with the Metropolitan Museum, told me. "The most extreme I know of is on a legal document of the Ramesside Period. It reads, 'As for the one who will violate it, he shall be seized for Amun-Ra. He shall be for the flame of Sekhmet. He is an enemy of Osiris, lord of Abydos, and so is his son, for ever and ever. May donkeys fuck him, may donkeys fuck his wife, may his wife fuck his son.'"

Some scholars today, looking back over the past two hundred years of archaeological activity, think that a curse might have been a good idea: most of the archaeology done in the Valley has been indistinguishable from looting. Until the 1960s, those who had concessions to excavate there were allowed to keep a percentage of the spoils as "payment" for their work. In the fever of the treasure hunt, tombs were emptied without anyone bothering to photograph the objects found or to record their positions in situ, or even to note which tomb they came from. Items that had no market value were trashed. Wilkinson, the man who gave the tombs their numbers, burned three-thousand-year-old wooden coffins and artifacts to heat his house. Murals and reliefs were chopped out of walls. At dinner parties, the American lawyer Theodore M. Davis, who financed many digs in the Valley, used to tear up necklaces woven of ancient flowers and fabric to show how strong they were after three thousand years in a tomb. Pyramids were blasted open with explosives, and one tomb door was bashed in with a battering ram. Even Carter never published a proper scientific report on Tut's tomb. It is only in the last twenty-five years that real archaeology has come to Egypt, and KV5 will be one of the first tombs in the Valley of the Kings to be entirely excavated and documented according to proper archaeological techniques.

Fortunately, other great archaeological projects remain to be carried out with the new techniques. The Theban Necropolis is believed to contain between four thousand and five thousand tombs, of which only four hundred have been given numbers. More than half of the royal tombs in the Valley of the Kings have not been fully excavated, and of these only five have been properly documented. There are mysterious blocked passageways, hollow floors, chambers packed with debris, and caved-in rooms. King Tut's was by no means the last undiscovered pharaonic tomb in Egypt. In the New Kingdom alone, the tombs of Amosis, Amenhotep I, Tuthmosis II, and Ramesses VIII have never been identified. The site of the burial ground for the pharaohs of the entire Twenty-first Dynasty is unknown. And the richness and size of KV5 offer the tantalizing suggestion that other princely tombs of its kind are lying undiscovered beneath the Egyptian sands; Ramesses would surely not have been the only pharaoh to bury his sons in such style.

WORK AT KV5 in the fall season proceeds from six thirty in the morning until one thirty in the afternoon. Every day, to get to KV5 from my hotel in Luxor, I cross the Nile on the public ferry, riding with a great mass of fellaheen—men carrying goats slung around their necks, children lugging sacks of eggplants, old men squatting in their djellabas and smoking cigarettes or eating *leb* nuts—while the ancient diesel boat wheezes and blubs across the river. I am usually on the river in time to catch the sun rising over the shattered columns of Luxor Temple, along the riverbank. The Nile is still magical—crowded with feluccas, lined with date palms, and bearing on its current many clumps of blooming water hyacinths.

The ferry empties its crowds into a chaos of taxis, camels, donkeys, children begging for baksheesh, and hopeful guides greeting every tourist with a hearty "Welcome to Egypt!" In contrast to the grand hotels and boulevards of Luxor, the west bank consists of clusters of mud villages scattered among impossibly green fields of cane and clover, where the air is heavy with smoke and the droning prayers of the muezzin. Disembarkation is followed by a harrowing high-speed taxi ride to the Valley, the driver weaving past donkey carts and herds of goats, his sweaty fist pounding the horn.

On the first day of my visit, I find Kent Weeks sitting in a green canvas tent at the entrance to KV5 and trying to fit together pieces of a human skull. It is a cool Saturday morning in November. From the outside, KV5 looks like all the other tombs—a mere doorway in a hillside. Workmen in a bucket brigade are passing baskets filled with dirt out of the tomb's entrance and dumping them in a nearby pile, on which two men are squatting and sifting through the debris with small gardening tools. "Hmm," Weeks says, still fiddling with the skull. "I had this together a moment ago. You'll have to wait for our expert. He can put it together just like that." He snaps his fingers.

"Whose skull is it?"

"One of Ramesses' sons, I hope. The brown staining on it—here—shows that it might have come from a mummified body. We'll eventually do DNA comparisons with Ramesses and other members of his family."

Relaxing in the tent, Weeks does not cut the dapper, pugnacious figure of a Howard Carter, nor does he resemble the sickly, elegant

Lord Carnarvon in waistcoat and watch chain. But because he is the first person to have made a major discovery in the Valley of the Kings since Carter, he is surely in their class. At fifty-four, he is handsome and fit, his ruddy face peering at the world through thick square glasses underneath a Tilley hat. His once crisp shirt and khakis look like hell after an hour in the tomb's stifling atmosphere, and his Timberland shoes have reached a state of indescribable lividity from tomb dust.

Weeks has the smug air of a man who is doing the most interesting thing he could possibly do in life. He launches into his subject with such enthusiasm that one's first impulse is to flee. But as he settles back in his rickety chair with the skull in one hand and a glass of *yansoon* tea in the other, and yarns on about lost tombs, crazy Egyptologists, graverobbers, jackal-headed gods, mummies, secret passageways, and the mysteries of the Underworld, you begin to succumb. His conversation is laced with obscene sallies delivered with a schoolboy's relish, and you can tell he has not been to any gender-sensitivity training seminars. He can be disconcertingly blunt. He characterized one archaeologist as "ineffectual, ridiculously inept, and a wonderful source of comic relief," another as "a raving psychopath," and a third as "a dork, totally off the wall." When I asked if KV5 would prove that Ramesses was the Biblical Pharaoh, he responded with irritation: "I can almost guarantee you that we will *not* find anything in KV5 bearing on the Exodus question. All the speculation in the press assumed there *was* an exodus and that it was described accurately in the Bible. I don't believe it. There may have been Israelites in Egypt, but I sincerely doubt whether Exodus is an exact account of what occurred. At least I *hope* it wasn't—with the Lord striking down the firstborn of Egypt and turning the rivers to blood."

His is a rarefied profession: there are only about four hundred Egyptologists in the world, and only a fraction of them are archaeologists. (Most are art historians and philologists.) Egyptology is a difficult profession to break into; in a good year, there might be two job openings in the United States. It is the kind of field where the untimely death of a tenured figure sets the photocopying machines running all night.

"From the age of eight, I had no doubt: I wanted to be an Egyptologist," Weeks told me. His parents—one a policeman, the other a medical librarian—did not try to steer him into a sensible profession, and a string of teachers encouraged his interest. When Weeks was in high school, in Longview, Washington, he met the Egyptologist Ahmed Fakhry in Seattle, and Fakhry was so charmed by the young man that he invited him to lunch and mapped out his college career.

In 1963, Weeks's senior year at the University of Washington, one of the most important events in the history of Egyptology took place. Because of the construction of the High Dam at Aswan, the rising waters of the Nile began to flood Nubia; they would soon inundate countless archaeological sites, including the incomparable temples of Abu Simbel. UNESCO and the Egyptian and Sudanese governments issued an international plea for help. Weeks immediately wrote to William Kelly Simpson, a prominent Egyptologist at Yale who was helping to coordinate the salvage project, and offered his services. He received plane tickets by return mail.

"The farthest I'd been away from home was Disneyland, and here I was going to Nubia," Weeks said. "The work had to be done fast: the lake waters were already rising. I got there and suddenly found myself being told, 'Take these eighty workmen and go dig that ancient village.' The nearest settlement was Wadi Halfa, ninety miles away. The first words of Arabic I learned were 'Dig no deeper' and 'Carry the buckets faster.'"

Weeks thereafter made a number of trips to Nubia, and just before he set out on them he invited along as artist a young woman he had met near the mummy case at the University of Washington museum—Susan Howe, a solemn college senior with red hair and a deadpan sense of humor.

"We lived on the river on an old rat-infested dahabeah," Susan told me. "My first night on the Nile, we were anchored directly in front of Abu Simbel, parked right in front of Ramesses' knees. It was all lit up, because work was on day and night." An emergency labor force was cutting the temple into enormous blocks and reassembling it on higher ground. "After five months, the beer ran out, the cigarettes ran out, the water was really hot, the temperature was a hundred and fifteen degrees in the shade, there were terrible windstorms. My parents were just *desperate* to know when I was coming home. But I thought, Ah! This is the life! It was so romantic. The workmen sang songs and clapped every morning when we arrived. So we wrote home and gave our parents ten days' notice that we were going to get married."

They have now been married twenty-nine years. Susan is the artist and illustrator for many of Kent's projects, and has also worked for other archaeologists in Egypt. She spends much of her day in front of KV5, in the green tent, wearing a scarf and peach-colored Keds, while she makes precise scale drawings of pottery and artifacts. In her spare time, she wanders around Gezira Bairat, painting exquisite watercolors of doorways and donkeys.

Weeks eventually returned to Washington to get his M.A., and in 1971 he received a Ph.D. from Yale; his dissertation dealt with ancient Egyptian anatomical terminology. He landed a plum job as a curator in the Metropolitan Museum's Egyptian Department. Two years later, bored by museum work, he quit and went back to Egypt, and was shortly offered the directorship of Chicago House, the University of Chicago's research center in Luxor. The Weekses have two children, whom they reared partly in Egypt, sending them to a local Luxor school. After four years at Chicago House, Weeks took a professorship at Berkeley, but again the lure of Egypt was too strong. In 1987, he renounced tenure at Berkeley, took a large pay cut, and went back to Egypt as a professor of Egyptology at the American University of Cairo, where he has been ever since.

While in Nubia, Weeks excavated an ancient working-class cemetery, pulling some seven thousand naturally desiccated bodies out of the ground. In a study of diet and health, he and a professor of orthodontics named James Harris X-rayed many of these bodies. Then Weeks and Harris persuaded the Egyptian government to allow them to X-ray the mummies of the pharaohs, by way of comparison. A team of physicians, orthodontists, and pathologists studied the royal X-rays, hoping to determine such things as age at death, cause of death, diet, and medical problems. They learned that there was surprisingly little difference between the two classes in diet and health.

One finding caused an uproar among Egyptologists. The medical team had been able to determine ages at death for most of the pharaohs, and in some cases these starkly contradicted the standard chronologies of the Egyptologists. The mystery was eventually solved when the team consulted additional ancient papyri, which told how, in the late New Kingdom, the high priests realized that many of the tombs in the Valley of the Kings had been robbed. To prevent further desecration, they gathered up almost all the royal mummies (missing only King Tut) and reburied them in two caches, both of which were discovered intact in the nineteenth century. "What we think happened is that the priests let the name dockets with some of the mummies fall off and put them back wrong," Weeks told me. It is also possible that the mix-up occurred when the mummies were moved down the river to Cairo in the nineteenth century.

The team members analyzed the craniofacial characteristics of each mummy and figured out which ones looked most like which others. (Most of the pharaohs were related.) By combining these findings with age-at-death information, they were able to restore six of the mummies' proper names.

Weeks's second project led directly to the discovery of KV5. In 1979, he began mapping the entire Theban Necropolis. After an overview, he started with the Valley of the Kings. No such map had ever been done before. (That explains how KV5 came to be found then lost several times in its history.) The Theban Mapping Project is to include the topography of the Valley and the three-dimensional placement of each tomb within the rock. The data are being computerized, and eventually Weeks will re-create the Valley on CD-ROM, which will allow a person to "fly" into any tomb and view in detail the murals and reliefs on its walls and ceilings.

Some Egyptologists I spoke with consider the mapping of the Theban Necropolis to be the most important archaeological project in Egypt, KV5 notwithstanding. A map of the Valley of the Kings is desperately needed. Some tombs are deteriorating rapidly, with murals cracking and falling to the floors, and ceilings, too, collapsing. Damage has been done by the opening of the tombs to outside air. (When Carter opened King Tut's tomb, he could actually hear "strange rustling, murmuring, whispering sounds" of objects as the new air began its insidious work of destruction. In other tombs, wooden objects turned into "cigar ash.") Greek, Roman, and early European tourists explored the tombs with burning torches—and even lived in some tombs—leaving an oily soot on the paintings. Rapid changes in temperature and humidity generated by the daily influx of modern-day tourists have caused even greater damage, some of it catastrophic.

The gravest danger of all comes from flooding. Most of the tombs are now wide open. Modern alterations in the topography, such as the raising of the valley floor in order to build paths for the tourists, have created a highway directing floodwaters straight into the mouths of tombs. A brief rain in November 1994 generated a small flash flood that tore through the Valley at thirty miles an hour and damaged several tombs. It burst into the tomb of Bay, a vizier of the New Kingdom, with such force that it churned through the decorated chambers and completely ruined them. Layers of debris in KV5 indicate that a major flash flood occurs about once every three hundred years. If such a flood occurred tomorrow, the Valley of the Kings could be largely destroyed.

There is no master plan for preserving the Valley. The most basic element in such a plan is the completion of Weeks's map. Only then can preservationists monitor changes in the tombs and begin channeling and redirecting floodwaters. For this reason, some archaeologists privately panicked when Weeks found KV5. "When I first heard about it," one told me, "I thought, Oh my God, that's it, Kent will never finish the mapping project."

Weeks promises that KV5 will not interfere with the Theban Mapping Project. "Having found the tomb, we've got an obligation to leave it in a good, stable, safe condition," he says. "And we have an obligation to publish. Public interest in KV5 has actually increased funding for the Theban Mapping Project."

AT NINE A.M., the workmen laboring in KV5—there are forty-two of them—begin to file out and perch in groups on the hillside, to eat a breakfast of bread, tomatoes, green onions, and a foul cheese called *misht*. Weeks rises from his chair, nods to me, and asks, "Are you ready?"

We descend a new wooden staircase into the mountain and enter Chamber 1, where we exchange our sun hats for hard hats. The room is small and only half cleared. Visible tendrils of humid, dusty air waft in from the dim recesses of the tomb. The first impression I have of the tomb is one of shocking devastation. The ceilings are shot through with cracks, and in places they have caved in, dropping automobile-size pieces of rock. A forest of screw jacks and timbers holds up what is left, and many of the cracks are plastered with "tell-tales"—small seals that show if any more movement of the rock occurs.

The reliefs in Chamber 1 are barely visible, a mere palimpsest of what were once superbly carved and painted scenes of Ramesses and his sons adoring the gods, and panels of hieroglyphics. Most of the damage here was the result of a leaky sewer pipe that was laid over the tomb about forty years ago from an old rest house in the Valley. The leak caused salt crystals to grow and eat away at the limestone walls. Here and there, however, one can still see traces of the original paint.

The decorations on the walls of the first two rooms show various sons being presented to the gods by Ramesses, in the classic Egyptian pose: head in profile, shoulders in frontal view, and torso in three-quarters view. There are also reliefs of tables laden with offerings of food for the gods, and hieroglyphic texts spelling out the names and titles of several sons and including the royal cartouche of Ramesses.

A doorway from Chamber 2 opens into Chamber 3—the Pillared Hall. It is filled with dirt and rock almost to the ceiling, giving one a simultaneous impression of grandeur and claustrophobia. Two narrow channels have been cut through the debris to allow for the passage of workmen. Many of the pillars are split and shattered, and only fragments of decorations remain—a few hieroglyphic characters, an upraised arm, part of a leg. Crazed light from several randomly placed bulbs throws shadows around the room.

I follow Weeks down one of the channels. "This room is in such dangerous condition that we decided not to clear it," he says. "We call this channel the Mubarak trench. It was dug so that President Mubarak could visit the tomb without having to creep around on his hands and knees." He laughs.

When we are halfway across the room, he points out the words "James Burton 1825" smoked on the ceiling with the flame of a candle: it represents the Englishman's farthest point of penetration. Not far away is another graffito—this one in hieratic, the cursive form of hieroglyphic writing. It reads "Year 19"—the nineteenth year of Ramesses' reign. "This date gives us a *terminus ante quem* for the presence of Ramesses' workmen in this chamber," Weeks says.

He stops at one of the massive pillars. "And here's a mystery," he says. "Fifteen of the pillars in this room were cut from the native rock, but this one is a fake. The rock was carefully cut away—you can see chisel marks on the ceiling—and then the pillar was rebuilt out of stone and plastered to look like the others. Why?" He gives the pillar a sly pat. "Was something very large moved in here?"

I follow Weeks to the end of the trench—the site of the doorway that he crawled through in February. The door has been cleared, and we descend a short wooden staircase to the bottom of the great central corridor. It is illuminated by a string of naked light bulbs, which cast a yellow glow through a pall of dust. The many doors lining both sides of the corridor are still blocked with debris, and the stone floor is covered with an inch of dust.

At the far end of the corridor, a hundred feet away, stands the mummiform statue of Osiris. It is carved from the native rock, and only its face is missing. Lit from below, the statue casts a dramatic shadow on the ceiling. I try to take notes, but my glasses have fogged up, and sweat is dripping onto my notebook, making the ink run off the page. I can only stand and blink.

Nothing in twenty years of writing about archaeology has prepared me for this great wrecked corridor chiseled out of the living rock, with rows of shattered doorways opening into darkness, and ending in the faceless mummy of Osiris. I feel like a trespasser, a voyeur, gazing into the sacred precincts of the dead. As I stare at the walls, patterns and lines begin to emerge from the shattered stone: ghostly figures and faint hieroglyphics; animal-headed gods performing mysterious rites. Through doorways I catch glimpses of more rooms and more doorways beyond. There is a presence of death in this wrecked tomb that goes beyond those who were buried here; it is the death of a civilization.

With most of the texts on the walls destroyed or still buried under debris, it is not yet possible to determine what function was served by the dozens of side chambers. Weeks feels it likely, however, that they were *not* burial chambers, because the doorways are too narrow to admit a sarcophagus. Instead, he speculates they were chapels where the Theban priests could make offerings to the dead sons. Because the tomb departs so radically from the standard design, it is impossible even to speculate what the mysterious Pillared Hall or many of the other antechambers were for.

Weeks proudly displays some reliefs on the walls, tracing with his hand the figure of Isis and her husband, Osiris, and pointing out the ibis-headed god Thoth. "Ah!" he cries. "And here is a *wonderful* figure of Anubis and Hathor!" Anubis is the jackal-headed god of mummification, and Hathor a goddess associated with the Theban Necropolis. These were scenes to help guide Ramesses' sons through the rituals, spells, and incantations that would insure them a safe journey through the realm of death. The reliefs are exceedingly difficult to see; Susan Weeks told me later that she has sometimes had to stare at a wall for long periods—days, even—before she could pick out the shadow of a design. She is now in the process of copying these fragmentary reliefs on Mylar film, to help experts who will attempt to reconstruct the entire wall sequence and its accompanying text, and so reveal to us the purpose of the room or the corridor. KV5 will only yield up its secrets slowly, and with great effort.

"Here's Ramesses and one of his sons," Weeks says, indicating two figures standing hand in hand. "But, alas, the name is gone. Very disappointing!" He charges off down the corridor, raising a trail of dust, and comes to a halt at the statue of Osiris, poking his glasses back up his sweating nose. "Look at this. Spectacular! A three-dimensional statue of Osiris is very rare. Most tombs depict him painted only. We dug around the base here trying to find the face, but instead we found a lovely offering of nineteen clay figs."

He makes a ninety-degree turn down the left transverse corridor, snaking around a cave-in. The corridor runs level for some distance and then plunges down a double staircase with a ramp in the middle, cut from the bedrock, and ends in a wall of bedrock. Along the sides of this corridor we have passed sixteen more partly blocked doors.

"Now, here is something new," Weeks says. "You're the first outsider to see this. I hoped that this staircase would lead to the burial chambers. This kind of ramp was usually built to slide the sarcophagi down. But look! The corridor just ends in a blank wall. Why in the world would they build a staircase and ramp going nowhere? So I decided to clear the two lowest side chambers. We just finished last week."

He ushers me into one of the rooms. There is no light; the room is large and very hot.

"They were empty," Weeks says.

"Too bad."

"Take a look at this floor."

"Nice." Floors do not particularly excite me.

"It happens to be the finest plastered floor in the Valley of the Kings. They went to enormous trouble with this floor, laying down three coats of plaster at different times, in different colors. Why?" He pauses. "Now stamp on the floor."

I thump the floor. There is a hollow reverberation that shakes not only the floor but the entire room. "Oh my God, there's something underneath there!" I exclaim.

"*Maybe*," Weeks says, a large smile gathering on his face. "Who knows? It could be a natural cavity or crack, or it might be a passageway to a lower level."

"You mean there might be sealed burial chambers below?"

Weeks smiles again. "Let's not get ahead of ourselves. Next June, we'll drill some test holes and do it properly."

We scramble back to the Osiris statue.

"Now I'm going to take you to our latest discovery," Weeks says. "This is intriguing. *Very* fascinating."

We make our way through several turns back to the Pillared Hall. Weeks leads me down the other trench, which ends at the southwest corner of the hall. Here, earlier in the month, the workmen discovered a buried doorway that opened onto a steep descending passageway, again packed solid with debris. The workmen have now cleared the passageway down some sixty feet, exposing twelve more side chambers, and are still at work.

We pause at the top of the newly excavated passageway. A dozen screw jacks with timbers hold up its cracked ceiling. The men have finished breakfast and are back at work, one man picking away at the wall of debris at the bottom of the passageway while another scoops the debris into a basket made out of old tires. A line of workmen then pass the basket up the corridor and out of the tomb.

"I've called this passageway 3A," Weeks says. He drops his voice. "The incredible thing is that this corridor is heading toward the tomb of Ramesses himself. If it connects, that will be extraordinary. No two tombs were ever deliberately connected. This tomb just gets curiouser and curiouser."

Ramesses' tomb, lying a hundred feet across the Valley, was also wrecked by flooding and is now being excavated by a French team. "I would dearly love to surprise them," Weeks says. "To pop out one day and say '*Bonjour! C'est moi!*' I'd love to beat the French into their own tomb."

I follow him down the newly discovered corridor, slipping and sliding on the pitched floor. "Of course," he shouts over his shoulder, "the sons might also be buried *underneath* their father! We clearly haven't found the burial chambers yet, and it is my profound hope that one way or another this passageway will take us there."

We come to the end, where the workmen are picking away at the massive wall of dirt that blocks the passage. The forty-two men can remove about nine tons of dirt a day.

At the bottom, Weeks introduces me to a tall, handsome Egyptian with a black mustache and wearing a baseball cap on backward. "This is Muhammad Mahmud," Weeks says. "One of the senior workmen."

I shake his hand. "What do you hope to find down here?"

"Something very nice, *inshallah*."

"What's in these side rooms?" I ask Weeks. All the doorways are blocked with dirt.

Weeks shrugs. "We haven't been in those rooms yet."

"Would it be possible . . ." I start to ask.

He grins. "You mean, would you like to be the first human being in three thousand years to enter a chamber in an ancient Egyptian tomb? Maybe Saturday."

As we are leaving the tomb, I am struck by the amount of work still unfinished. Weeks has managed to dig out only three rooms completely and clear eight others partway—leaving more than eighty rooms entirely untouched. What treasures lie under five or ten feet of debris in those rooms is anyone's guess. It will take from six to ten more years to clear and stabilize the tomb, and then many more years to publish the findings from it. As we emerge from the darkness, Weeks says, "I know what I'll be doing for the rest of my life."

ONE MORNING, I find a pudgy, bearded man sitting in the green tent and examining, Hamlet-like, the now assembled skull. He is the paleontologist Elwyn Simons, who has spent decades searching the sands of the Faiyum for primate ancestors of human beings. Susan Weeks once worked for him, and now he is a close friend of the couple, dropping in on occasion to look over bones from the tomb. Kent and Susan are both present, waiting to hear his opinions about the skull's sex. (Only DNA testing can confirm whether it's an actual son of Ramesses, of course.)

Simons rotates the skull, pursing his lips. "Probably a male, because it has fairly pronounced brow ridges," he says. "This"—he points to a hole punched in the top of the cranium—"was made post-mortem. You can tell because the edges are sharp and there are no suppressed fractures."

Simons laughs, and sets the skull down. "You can grind this up and put it in your soup, Kent."

When the laughter has died down, I venture that I didn't get the joke.

"In the Middle Ages, people filled bottles with powdered mummies and sold it as medicine," Simons explains.

"Or mummies were burned to power the railroad," Weeks adds. "I don't know how many miles you get per mummy, do you, Elwyn?"

While talk of mummies proceeds, a worker brings a tray of tea. Susan Weeks takes the skull away and puts another bone in front of Simons.

"That's the scapula of an artiodactyl. Probably a cow. The camel hadn't reached Egypt by the Nineteenth Dynasty."

The next item is a tooth.

"Artiodactyl again," he says, sipping his tea. "Goat or gazelle."

The identification process goes on.

The many animal bones found in KV5 were probably offerings for the dead: valley tombs often contained sacrificed bulls, mummified baboons, birds, and cats, as well as steaks and veal chops.

Suddenly, Muhammad appears at the mouth of the tomb. "Please, Dr. Kent," he says. And starts telling Weeks in Arabic that the workers have uncovered something for him to see. Weeks motions for me to follow him into the dim interior. We put on our hard hats and duck through the first chambers into Corridor 3A. A beautiful set of carved limestone steps has appeared where I saw only rubble a few days before. Weeks kneels and brushes the dirt away, excited about the fine workmanship.

Muhammad and Weeks go to inspect another area of the tomb, where fragments of painted and carved plaster are being uncovered. I stay to watch the workmen digging in 3A. After a while, they forget I am there and begin singing, handing the baskets up the long corridor, their bare feet white with dust. A dark hole begins to appear between the top of the debris and the ceiling. It looks as if one could crawl inside and perhaps look farther down the corridor.

"May I take a look in there?" I ask. One of the workmen hoists me up the wall of dirt, and I lie on my stomach and wriggle into the gap. I recall that archaeologists sometimes sent small boys into tombs through holes just like this.

Unfortunately, I am not a small boy, and in my eagerness I find myself thoroughly wedged. It is pitch-black, and I wonder why I thought this would be exciting.

"Pull me out!" I yell.

The Egyptians heave on my legs, and I come sliding down with a shower of dirt. After the laughter subsides, a skinny man named Nubie crawls into the hole. In a moment, he is back out, feet first. He cannot see anything; they need to dig more.

The workmen redouble their efforts, laughing, joking, and singing. Working in KV5 is a coveted job in the surrounding villages; Weeks pays his workmen four hundred Egyptian pounds a month (about a hundred and twenty-five dollars), four times what a junior inspector of antiquities makes and perhaps three times the average monthly income of an Egyptian family. Weeks is well liked by his Egyptian workers, and is constantly bombarded with dinner invitations from even his poorest laborers. While I was there, I attended three of these dinners. The flow of food was limitless, and the conversation competed with the bellowing of a water buffalo in the adjacent room or the braying of a donkey tethered at the door.

After the hole has been widened a bit, Nubie goes up again with a light and comes back down. There is great disappointment: it looks as though the passageway might come to an end. Another step is exposed in the staircase, along with a great deal of broken pottery. Weeks returns and examines the hole himself without comment.

AS THE WEEK goes by, more of Corridor 3A is cleared, foot by foot. The staircase in 3A levels out to a finely made floor, more evidence that the corridor merely ends in a small chamber. On Wednesday, however, Weeks emerges from the tomb smiling. "Come," he says.

The hole in 3A has now been enlarged to about two feet in diameter. I scramble up the dirt and peer inside with a light, choking on the dust. As before, the chiseled ceiling comes to an abrupt end, but below it lies what looks like a shattered door lintel.

"It's got to be a door," Weeks says, excited. "I'm afraid we're going to have to halt for the season at that doorway. We'll break through next June."

Later, outdoors, I find myself coughing up flecks of mud.

"Tomb cough," Weeks says cheerfully.

ON THURSDAY MORNING, Weeks is away on business, and I go down into the tomb with Susan. At the bottom of 3A, we stop to watch Ahmed Mahmud Hassan, the chief supervisor of the crew, sorting through some loose dirt at floor level. Suddenly, he straightens up, holding a perfect alabaster statuette of a mummy.

"Madame," he says, holding it out.

Susan begins to laugh. "Ahmed, that's beautiful. Did you get that at one of the souvenir stalls?"

"No," he says, "I just found it." He points to the spot. "Here."

She turns to me. "They once put a rubber cobra in here. Everyone was terrified, and Muhammad began beating it with a rock."

"Madame," Ahmed says. "Look, please." By now, he is laughing, too.

"I see it," Susan says. "I hope it wasn't too expensive."

"Madame, please."

Susan takes it, and there is a sudden silence. "It's real," she says quietly.

"This is what I was telling Madame," Ahmed says, still laughing.

Susan slowly turns it over in her hands.

"It's beautiful. Let's take it outside."

In the sunlight, the statuette glows. The head and shoulders still have clear traces of black paint, and the eyes look slightly crossed. It is an *ushabti*, a statuette that was buried only with the dead, meant to spare the deceased toil in the afterlife: whenever the deceased was called upon to do work, he would send the ushabti in his place.

That morning, the workmen also find in 3A a chunk of stone. Weeks hefts it. "This is very important," he says.

"How?"

"It's a piece of a sidewall of a sarcophagus that probably held one of Ramesses' sons. It's made out of serpentine, a valuable stone in ancient Egypt." He pulls out a tape measure and marks off the thickness of the rim. "It's eight-point-five centimeters, which, doubled, gives seventeen centimeters. Add to that the width of an average pair of human shoulders, and perhaps an inner coffin, and you could not have fitted this sarcophagus through any door to any of the sixty side chambers in that tomb." He pauses. "So, you see, this piece of stone is one more piece of evidence that we have yet to find the burial chambers."

Setting the stone down with a thud on a specimen mat, he dabs his forehead. He proceeds to lay out a theory about KV5. Ramesses had an accomplished son named Khaemwaset, who became the high priest of an important cult that worshipped a god represented by a sacred bull. In Year 16, Khaemwaset began construction of the Serapeum, a vast catacomb for the bulls, in Saqqara. The original design of the Serapeum is the only one that remotely resembles KV5's layout, and it might have been started about the same time. In the Serapeum, there are two levels: an upper level of offering chapels and a lower level for burials. "But," Weeks adds, throwing open his arms, "until we find the burial chambers it's *all* speculation."

ON FRIDAY, BRUCE Ludwig arrives—a great bear of a man with white hair and a white beard. Dressed like an explorer, he is lugging a backpack full of French wine for the team.

Unlike Lord Carnarvon and other wealthy patrons who funded digs in the Valley of the Kings, Ludwig is a self-made man. His father owned a grocery store in South Dakota called Ludwig's Superette. Bruce Ludwig made his money in California real estate and is now a partner in a firm managing four billion dollars in pension funds.

He has been supporting Weeks and the Theban Mapping Project for twelve years.

Over the past three, he has sunk a good deal of his own money into the project and has raised much more among his friends. Nevertheless, the cost of excavation continually threatens to outstrip the funds at hand. "The thing is, it doesn't take a Rockefeller or a Getty to be involved," he told me over a bottle of Château Lynch-Bages. "What I like to do is show other successful people that it's just hugely rewarding. Buildings crumble and fall down, but when you put something in the books, it's there forever."

Ludwig's long-term support paid off last February, when he became one of the first people to crawl into the recesses of KV5. There may be better moments to come. "When I discover that door covered with unbroken Nineteenth Dynasty seals," Weeks told me, joking, "you bet I'll hold off until Bruce can get here."

SATURDAY, THE WORKERS' taxi picks the Weekses and me up before sunrise and then winds through a number of small villages, collecting workers as it goes along.

The season is drawing to a close, and Susan and Kent Weeks are both subdued. In the last few weeks, the probable number of rooms in the tomb has increased from sixty-seven to ninety-two, with no end in sight. Everyone is frustrated at having to lock up the tomb now, leaving the doorway at the bottom of 3A sealed, the plaster floor unplumbed, the burial chambers still not found, and so many rooms unexcavated.

Weeks plans to tour the United States lecturing and raising more funds. He estimates he will need a quarter of a million dollars per year for the indefinite future in order to do the job right.

As we drive alongside sugarcane fields, the sun boils up over the Nile Valley through a screen of palms, burning into the mists lying on the fields. We pass a man driving a donkey cart loaded with tires, and whizz by the Colossi of Memnon, two enormous wrecked statues standing alone in a farmer's field. The taxi begins the climb to a village once famous for tomb robbing, some of whose younger residents now work for Weeks. The houses are completely surrounded by the black pits of tombs. The fragrant smell of dung fires drifts through the rocky streets.

Along the way, I talk with Ahmed, the chief supervisor. A young man with a handsome, aristocratic face, who comes from a prominent family in Gezira Bairat, he has worked for Weeks for about eight years. I asked him how he feels about working in the tomb.

Ahmed thinks for a moment, then says, "I forget myself in this tomb. It is so vast inside."

"How so?"

"I feel at home there. I know this thing. I can't express the feeling, but it's not so strange for me to be in this tomb. I feel something in there about myself. I am descended from these people who built this tomb. I can feel their blood is in me."

When we arrive in the Valley of the Kings, an inspector unlocks the metal gate in front of the tomb, and the workers file in, with Weeks leading the way. I wait outside to watch the sunrise. The tourists have not yet arrived, and if you screen out some signs you can imagine the Valley as it might have appeared when the pharaohs were buried here three thousand years ago. (The vendors and rest house have been removed.) As dawn strikes el-Qurn and invades the upper reaches of the canyon walls, a soft, peach-colored light fills the air. The encircling cliffs lock out the sounds of the world; the black doorways of the tombs are like dead eyes staring out; and one of the guard huts of the ancient priests can still be seen perched at the cliff edge. The whole valley becomes a slowly changing play of light and color, mountain and sky, unfolding in absolute stillness. I am given a brief, shivery insight into the sacredness of this landscape.

At seven, the tourists begin to arrive, and the spell is dispersed. The Valley rumbles to life with the grinding of diesel engines, the frantic expostulations of vendors, and the shouting of guides leading groups of tourists. KV5 is the first tomb in the Valley, and the tourists begin gathering at the rope, pointing and taking pictures, while the guides impart the most preposterous misinformation about the tomb: that Ramesses had four hundred sons by only two wives, that there are eight hundred rooms in the tomb, that the greedy Americans are digging for gold but won't find any. Two thousand tourists a day stand outside the entrance to KV5.

I go inside and find Weeks in 3A, supervising the placement of more screw jacks and timbers. When he has finished, he turns to me. "You ready?" He points to the lowest room in 3A. "This looks like a good one for you to explore."

One of the workmen clears away a hole at the top of the blocked door for me to crawl through, and then Muhammad gives me a leg up. I shove a caged light bulb into the hole ahead of me and wriggle through. I can barely fit.

In a moment, I am inside. I sit up and look around, the light throwing my distorted shadow against the wall. There is three feet of space between the top of the debris and the ceiling, just enough for me to crawl around on my hands and knees. The room is about nine feet square, the walls finely chiseled from the bedrock. Coils of dust drift past the light. The air is just breathable.

I run my fingers along the ancient chisel marks, which are as fresh as if they were made yesterday, and I think of the workmen who carved out this room, three millennia ago. Their only source of light would have been the dull illumination from wicks burning in a bowl of oil salted to reduce smoke. There was no way to tell the passage of time in the tomb: the wicks were cut to last eight hours, and when they guttered it meant that the day's work was done. The tombs were carved from the ceiling downward, the workers whacking off flakes of limestone with flint choppers, and then finishing the walls and ceilings with copper chisels and sandstone abrasive. Crouching in the hot stone chamber, I suddenly get a powerful sense of the enormous religious faith of the Egyptians. Nothing less could have motivated an entire society to pound these tombs out of rock.

Much of the Egyptian religion remains a mystery to us. It is full of contradictions, inexplicable rituals, and impenetrable texts. Amid the complexity, one simple fact stands out: it was a great human bargain with death. Almost everything that ancient Egypt has left us—the pyramids, the tombs, the temples—represents an attempt to overcome that awful mystery at the center of all our lives.

A shout brings me back to my senses.

"Find anything?" Weeks calls out.

"The room's empty," I say. "There's nothing in here but dust."

UPDATE

Kent and Susan Weeks moved to a houseboat on the Nile, where they lived while continuing to excavate KV5 and pursue the Theban Mapping Project. Kent published a well-received book about the discovery, The Lost Tomb, *in 1998. In 2009, after returning from a lecture at the Mummification Museum in Luxor, Susan apparently fell off the houseboat and drowned.*

Further excavations of KV5 continued to reveal dozens more rooms and corridors. So far 120 rooms have been uncovered, but still blocked tunnels and rooms suggest that the tomb probably contains at least 150 chambers, making it the largest tomb in all of Egypt in terms of numbers of rooms. The burial chambers, however, have not yet been found. As of today, almost thirty years later, less than ten percent of the tomb has been cleared. More human remains were found, including skull fragments that were identified as the remains of Amun-her-khepeshef, the firstborn son of Ramesses and Queen Nefertari, as well as canopic jars labeled with his name that contained mummified organs. The mysterious hollow floor, however, has not yet been investigated. The tomb is so vast and so precarious that it will be many more decades before it is fully excavated, if ever.

AN EXTRA ADVENTURE

IN SEARCH OF THE SEVEN CITIES OF GOLD

WE FINALLY LAID eyes on Hawikuh, the first of the Seven Cities of Gold, on May 23, 1989. We were sitting in plastic chairs in front of a Zuni shepherd's cabin, drinking coffee by the glow of a kerosene lantern. The shepherd, Lincoln Harker, had snapped on a transistor radio—our arrival being a festive occasion—and we were listening to the Doors. The music crackled and hissed, as if coming to us from a vast distance.

We were facing east, looking out over an empty plain rimmed by mesas. The sharp cooling of the air brought a gunpowdery smell of dust into our nostrils.

We told Lincoln that we were following the route of the Spanish explorer Coronado on his search for the Seven Cities of Gold. Tomorrow, we said, we hoped to reach Hawikuh. He listened, head sunk on chest, cupping his coffee mug in both hands. When we finished he leaned his chair against the cabin with a thump and pointed into the dark.

"That is Hawikuh," he said.

That's when we saw it: an unnaturally barren patch of ground on the shoulder of a hill.

Originally published in *Smithsonian Magazine* in 1990 as "On the Trail of Coronado and His Cities of Gold."

We had started our journey six weeks and 600 miles before. Our plan had been to follow, on horseback, part of the route that Francisco Vásquez de Coronado and his army took in 1540, while looking for the Seven Cities of Gold. We would ride cross-country, not following roads or trails, and would camp in the desert, much as Coronado and his men did 450 years ago. We would pack our food—mostly flour, beans and rice—on horseback, and find water and grass where we could. Our only concession to the modern age was a set of 175 U.S. Geological Survey topographical maps on which we had plotted our route.

Coronado's expedition was one of the most remarkable, and in some ways one of the most absurd, in the annals of exploration. In 1539 a friar named Marcos de Niza returned to Mexico City with a story that electrified the New World. Far to the north he had discovered "the greatest thing in the world"—a city, "larger than the city of Mexico," with buildings up to ten stories high. The surrounding kingdom, called Cibola, was the "best and greatest that had ever been heard of" and consisted of seven equally magnificent cities. Fray Marcos had hastened back to New Spain with the news, and soon everyone was talking about the Seven Cities of Gold. It was another Inca or Aztec empire, waiting to be conquered.

The Viceroy of New Spain awarded the plum job of conquest to the rising young nobleman Francisco Vásquez de Coronado, who quickly assembled a great army of nearly 1,500 men as well as slaves, more than 1,000 horses, and herds of sheep and cattle. Most of the men were in their teens and 20s. Coronado himself was scarcely 30 years old.

This army departed Compostela in February 1540. The Spaniards reached Culiacán around Easter. Coronado then went ahead with 100 lightly equipped men and Fray Marcos as the guide. In late May, Coronado and his advance guard finally entered what is now the United States, in southeastern Arizona. The discovery of the Southwest had begun.

We planned to follow Coronado's route from where he crossed into the United States to the ruins of Pecos Pueblo (now the Pecos National Monument)—a ground distance of over a thousand miles. Our journey would traverse some of the wildest and most spectacular desert and mountain country in the West—country so remote that it had changed little since Coronado's day. Although we were no admirers of Coronado, we hoped we might be able to recapture, in a small way, what the original discovery of the Southwest might have been like.

In mid-April, we set off at the Arizona-Mexico border, where the San Pedro River entered the United States, flowing under a five-strand barbed-wire fence. The river was low—little more than a braided stream sliding over a buckskin-colored bed of sand. Unbeknownst to us, Arizona was edging into one of its worst droughts in half a century.

I knew very little about horses and packing, being (as they say) a damn Yankee from Boston. My partner Walter Nelson, a photographer, knew a little more, having grown up near the Red River in Texas, but both of us were greenhorns of the rankest kind.

The trip started well. After being dropped off we camped a half-mile from the border at a cattle tank brimming with cold water. We went to sleep listening to the clanking, sucking, trilling sound of an old windmill, which reminded me of some Rube Goldberg contraption running on high.

When we woke at 4:45, my bedroll was crackling with hoarfrost. The sun rose at 6 and by 9:50 the temperature was close to 100 degrees, having spent all of 40 minutes climbing through the comfort zone. As we saddled up, the cottonwoods suddenly loosed their cotton, which swirled about us like snow.

Our high spirits did not last long. The brush and deadfall along the river stopped us cold, while the river bottom itself shivered with quicksand. The lead rope of my packhorse wedged up under my riding horse's tail, and I was bucked halfway across the river. Hoping that the floodplain might be easier, we struggled up a steep cutbank and forced our way through a thicket of mesquite to an immense field of yellow grass that was as tall as our horses' eyes. Almost immediately my horse tumbled into a five-foot-deep hole. I leapt off and the animal clambered out, badly frightened but unhurt. We dismounted and led our horses, sometimes having to feel our way along, as the field was riddled with treacherous holes rendered invisible by the blinding sun and grass.

In less than an hour we had to stop and repack our loads twice. Packing a horse properly is an operation that requires great skill and painstaking care, as well as powerful lungs and an inexhaustible supply of curses. It would take us weeks to master this delicate art.

We camped that night back down by the river. In a hard day's riding we had barely covered three and a half miles. As I lay in my bedroll, scratched, sore and hungry, two thoughts ran through my mind. The first was that we were going to be lucky to average 12 miles a day, and as a consequence the trip would take ten weeks or more, not the four to six I had estimated. The second was that our chances of finishing this trip without killing ourselves and our horses seemed slim indeed. The problem was that Walter and I had already spent $8,000 of other people's money to make this trip; we simply could not quit.

I was terrified.

That night migrants crossing the border from Mexico came up the river, pursued by the Border Patrol, and two of Walter's horses broke their hobbles and went galloping off in the dark. I huddled in my bag, pretending to be asleep while Walter hunted them down.

For its first 35 miles, the San Pedro flows through a wilderness of quicksand, cactus, cutbanks, sudden arroyos and brutal mesquite. We had to wear chaps and our heaviest clothing in the 100-degree heat, and every evening our horses' legs and our faces were scored and bloody from the mesquite thorns.

For the first few days, Walter, who was as grizzled a Texan as you will find, complained about my appearance, saying that with my new hat and sunglasses I looked like a "dern New Yorker on vacation." When we made camp on the fifth day, he came over and peered into my face. "You sure don't look green anymore," he said, "you're the dirtiest s.o.b. I think I've ever seen," and went off laughing. It was a badge of honor I would, unfortunately, wear for the rest of the trip.

We rode through landscapes of almost surreal beauty. Once we stopped for lunch in a deep stand of cottonwood and surprised a nesting colony of great blue herons: they circled and cried over our heads. Later we flushed a family of javelinas and sent them snorting and grunting into a field of tumbleweed. Other images are imprinted on my mind: blood-red arroyos slicing through curtains of green mesquite; distant blue mountains seen through a screen of cottonwood; an antelope bounding through grass as white as snow. Every evening, while the crowns of the trees filled with golden light, Walter and I would whoop and roll about in the shallow river, letting the cool water wash off the dust and sweat of the day. Every morning we rose in the freezing dark, and lit a fire to thaw ourselves out and make "cowboy" coffee. (Recipe: Three fistfuls of grounds in pot of bad water. Boil like hell until mixture is opaque and soupy. Drink or eat with spoon.)

Before the trip, we had worried about getting saddle-sore. Oddly enough, after five days everything *but* our rear ends began to hurt. On the first day, one of Walter's horses had stepped on my little toe, breaking it. It had swollen up like a Vienna sausage, and then I had left my cowboy boots in the sun for a few hours while taking a nap, shrinking them dreadfully. Wedging the broken toe into the front of the tight boot every morning became a perfect agony.

We spent most of our day, it seemed, tying and untying knots as hard as teak and gripping ropes leading to recalcitrant horses. Once I counted eight scabs, cuts and rope burns on my right hand and 14 on my left. In addition, all ten fingernails had either been ripped or chipped down to the quick.

Fortunately, in our medicine chest we found a tube of a substance called Bag Balm. Billed as an "Antiseptic Emollient Treatment for Horses & Cattle, Other Livestock, Small Animals," this marvelous embrocation was for "Hoofs, Body & Legs, Udder & Teats." It was also ideal for sore hands and butts. During the course of our journey we went through three large tubes of the stuff, very little of which was applied to the horses.

On the eighth day, I saw, across the river, the end of a dirt track leading into the little town of St. David. I nearly wept; a road never looked so good. In my eagerness to reach it I mired one of my horses, but managed to free it in a frenzy of shouting, amid flying clots of black mud. When we finally entered St. David, somebody called the police.

For years, scholars have argued about where Coronado traveled in Arizona. Two broad possibilities have been proposed: a western route up through east-central Arizona and an eastern route that roughly follows the modern Arizona-New Mexico border northward. For both the western and eastern route, two major alternatives have been suggested. We had chosen to follow the western route and to retrace, at least in part, its two possible trails.

We therefore left the river 15 miles above St. David and headed into the desert, following the western subroute most favored by experts on the Coronado period. Several days of riding brought us into a high, lonely country called the Mesas, lying between two rugged mountain chains. All mesquite and cactus vanished, leaving nothing but rolling grassy hills the color and texture of fine chamois. Later we saw a vast herd of black-tailed deer, hundreds of them covering an entire hillside, rippling and flowing like water.

Then we swung northwest to intercept the Gila River and rejoined the other major western route. Beyond the Gila the mountains began in earnest, wave after wave of them, finally culminating in the Mogollon Rim. By the time Coronado and his advance guard reached this point, they had begun to suspect that something was amiss. The good friar's glowing descriptions of the countryside—the fine pastures, flowing waters and easy trails—did not correspond to the cactus-choked plains, crack-bottomed arroyos and savage mountains they were struggling through.

We would soon be entering what the expedition chroniclers termed the great *despoblado*—a word denoting an "uninhabited area" but implying something more akin to a desolate wilderness. This was the mountainous country leading up to and beyond the Mogollon Rim, a high escarpment that stretches from western New Mexico nearly halfway across Arizona.

While passing through Globe, Arizona, we met a horseman named Mark Shellenberger, who put us up for the night and mapped out for us an alternative to the route we had planned. His route would shave 50 miles off our trip and, he pointed out, had the distinct advantage of not killing us in the process.

WE GOT OUR first glimpse of the Salt River on May 9, nearly a month into the trip; a shining loop of water curling through a deep gorge. Beyond rose layer upon layer of blue mountains as far as the eye could see—the heart of the great *despoblado*. We guessed that the farthest blue line must be the Rim itself; later we would realize it was merely Board Tree Saddle, not even halfway across Rim country. Gazing over this restless sea of mountains, I felt a shiver creep up my spine.

We camped next to the river. While bathing, I was surprised by a thin, ragged man with long hair and an enormous black beard. He wanted to know what the hell I was doing in his swimming hole. "You seen any rattlers?" he added. I said no. "Good," he said, and poked his stick through some old roots and weeds around the pool. When he was satisfied there were no snakes, he went for a dip.

His name was Rod, and he told me he'd been shot up in Vietnam and lived in a tent a quarter-mile upriver. He said he didn't "mix too good with civilization" and that he hunted rattlesnakes for a living, making hatbands and belts from their skins. I had never met a man so terrified of rattlesnakes. "Someday they're gonna git me," he said. "I've killed too many now." That evening, he came to dinner carrying a mess of catfish. Up to that point our breakfasts, lunches and dinners had consisted of what we called the Coronado Plan: boiled oatmeal in the morning; nuts and dried fruit for lunch; and a single pot of beans, rice and quinoa (a grain grown by the Incas, high in protein) with biscuits on the side, for dinner. We stuffed ourselves with such vehemence on the greasy fried catfish that Rod asked us if we were all right.

———

THE NEXT DAY Rod showed us a place to ford, hollering directions from the bank. The Salt, even at its low state because of the drought, was full of deep channels and sudden holes that would drown our packhorses (a packed horse will sink like a stone).

The ascent of the hogback ridge was the most frightening journey of our trip, through steep rocks and plunging cliffs choked with loose rock, cholla, yucca and saguaro cactus. High up on the ridge, with Walter leading, we rounded a shoulder. Walter stopped, and in an unnaturally calm voice he told me to turn around and go back. The trail ended in a chasm and his two horses were standing on a narrow ledge; with a sudden chill I realized that Walter was trapped. I did not see how he could possibly turn both his horses around. He couldn't even dismount, as that would simply have carried him over the side of the cliff.

There was a long silence. "Okay, horses, go!" he said and turned the reins into empty space. I looked away, but when the expected whinny and crash of falling horses did not occur, I turned back. Walter was riding toward me with a big grin. "Just gotta trust 'em," he said, patting his horse, Pedernal.

We stopped for lunch near the top, exhausted and shaking from stress. The horses were in a similar state and we spent 15 minutes pulling cactus spines out of their legs. From where we were, we could see the Salt River crawling through its gorge 1,000 feet below us, shut in by sheer cliffs on all sides. "Somehow," I said, "I don't think Coronado rode down that river." The disputed portion of Coronado's western trail ends near the junction of the Little Colorado and Zuni rivers, about 250 miles from where we stood at the Salt. At the Little Colorado we would pick up his actual trail again and ride it for the next 350 miles to Pecos.

We made slow progress through the Rim country. As we gained altitude, the mountainous desert gave way to a deep pine forest, filled with dripping ferns. One morning I climbed a hill to look for a missing horse. At the top the clouds suddenly parted, and there, rising like a 1,000-foot wall, was the Rim itself, covered with a layer of fresh, sparkling snow.

The Mogollon Rim marks the southern boundary of the vast Colorado Plateau. As we pushed northeast, the pines and lush grass gave way to cedars, and then we rode into vast treeless immensities of sagebrush, red rimrock and blue sky. Below the Rim it had taken a supreme effort just to make ten miles in a day; now we would easily log 20 or 25.

In late May we hit the junction of the Little Colorado and Zuni rivers, some 60 miles from the Seven Cities of Gold, and camped at the base of the Stinking Springs Mountains. We cooked by burning cow chips, and the water in our canteens got so hot in the blazing sun that it seared our lips when we drank. One night I woke suddenly and found myself staring into a boundless pasture of stars; for a moment I panicked, thinking I was falling upward into space.

Early in the trip we had realized that if our horses were to survive, we had to take better care of them than we did of ourselves. By this time everything revolved around the horses. The horses determined our route, where we camped, where we stopped for lunch and sometimes whether we went anywhere at all. As the horses lost weight, we became obsessed with finding them good grazing. Even so, the grazing was usually so sparse that we had to hobble the horses, to allow them to wander far and wide looking for grass. They would often travel two or three miles during the night, and we would have to spend hours tracking them the next morning. We would then ride them bareback into camp, singing at the top of our lungs.

SOMEWHERE EAST OF the Stinking Springs Mountains the Spanish first encountered the Zuni: from the top of a mesa, two Indians appeared and vanished. The next day, a group of Indians gathered above the Spanish camp and hollered and shouted, so startling the men that, in their panic, many saddled their horses backward. A day later, Coronado's haggard party topped a low rise and found themselves gazing on the first of the legendary Seven Cities of Gold.

One of the soldiers would later recall that moment: "Such were the curses that some hurled at Fray Marcos," he wrote, "that I pray God may protect him from them." Instead of a magnificent, shining city-fortress, the disappointed men saw only "a little, crowded village, looking as if it had been crumpled all up together." It was the Zuni Indian village of Hawikuh.

They found the province up in arms. The Spanish recited the usual bombastic oration to the effect that if the Indians would merely submit to the "protection" of the king of Spain and take Christ as their savior, they would be unharmed. It is unclear whether the Zuni understood any of this, but at any rate they quickly provided a logical answer: a shower of arrows. The fight lasted barely an hour. With the advantage of harquebuses, crossbows and armor, the Spanish routed the Indians.

But we encountered no Indians—and indeed no human beings—until we crossed the New Mexico border and saw Lincoln Harker standing in the doorway of his sheep camp, his hands raised in greeting. The plain of Hawikuh was empty, the city of gold in ruins. We had decided to break the trip at Zuni and finish in August after the start of the summer rains, when the grazing would be better.

The morning after our arrival at the sheep camp, we set off for Hawikuh. From the top of the ruin we had a sweeping view of the surrounding countryside: a wilderness of buttes, mesas, plains and hard-etched horizons. In the entire landscape there was no sign of human life except a lone Indian shepherd driving a flock of sheep across the plain, a little black dog trotting at his heels. I stretched out and rested my head on a chiseled block of stone. In the distance, I could hear the tinkling of sheep bells and the baaing of sheep, rising and falling on the wind. Then I heard a new sound, a low rumble, and an almost invisible jet left a snowy contrail across a field of blue; the sun turned once across its aluminum skin just before it dipped below the horizon. I covered my face with my hat and drifted off to sleep.

During the months following the defeat of Hawikuh, Coronado settled in at Zuni and sent exploring parties east and west. During this time a delegation of Indians arrived from a powerful pueblo called Pecos, bearing gifts for the "strange people, bold men" they had heard about. Coronado sent one of his captains, a man named Alvarado, back to Pecos with the Indians.

The old Indian trade route from Zuni to Pecos, which Alvarado followed, was the trail we would take (Coronado himself went by a different route later). The first thing we had to do was find three new horses, as the horses I had borrowed for myself for the first part of the trip were no longer available and one of Walter's horses was sick. I decided we should take a Spanish Barb along. This is the breed that many feel may have been the "horse of the Conquest," the horse that Coronado used.

Luckily, one of the top breeders of these horses was located in Santa Fe. I telephoned and asked if I could borrow one. After the initial shock wore off, they decided to lend me their prize brood mare—a magnificent dun. The horse was worth $10,000, being the last breeding mare in a famous Barb bloodline. Her name was Engwahela.

One of the stable's owners, Roeliff Annon, explained why they lent me the horse. "We are working to restore the horse of the Conquest. We'd made a lot of claims for what this breed could do, and I thought this was a good opportunity to put some of these claims to the test." I also bought a dirty-white Appaloosa named Wilbur, and a chestnut quarter horse named Redbone.

We resumed the trip in early August. A friend at Zuni, Edward Wemytewa, invited us to set up camp on his land. We rode out of Zuni on August 7 and almost immediately ran into trouble. The first night we camped behind Dowa Y'alanne, the sacred mesa of the Zuni, in a small meadow ringed by deep arroyos and badlands. The camp had no water, but we expected to find some the next morning. That night, in an unaccountable lapse of judgment, we let Engwahela graze free, expecting that she would not leave the three geldings. The next morning she was gone. "She can't be far," Walter said cheerfully as he set off following her tracks. "Make me a good breakfast."

I waited. The sun climbed in the sky and the heat became nearly intolerable. By noon I realized that I had better water the other horses, so I saddled up Redbone and prepared to take them to a hidden spring about four miles from our camp. I wrapped our last canteen around the saddle horn, and put up Redbone's reins for a second to remove Wilbur's hobbles and untie Pedernal.

Suddenly Redbone, who had been standing stock-still for the past four hours, decided to start trotting off, with Wilbur at his heels. I had committed the grievous sin of leaving two horses unguarded, even for a moment. Now I was going to pay for it. Leading Pedernal by his rope, I jogged after them, cooing and whispering about oats and grain and all the wonderful things they would have if they would only stop. They were not impressed. They looked back and broke into a gallop. In a moment of absolute despair, I threw myself on Pedernal bareback, and with only a lead rope to control him jabbed my heels into his flank.

In what can only be termed an explosion, he took off. He raced after his departing friends, leaping arroyos, plunging down banks and dodging piñon trees; in vain I hollered and tried to pull his head around with the lead rope. It was no use. I couldn't stay on him. I went flying, and slammed into a soft bank of dirt. By the time I figured out I wasn't dead and struggled to my feet, the horses were gone—carrying away the last of my water. In the direction they were headed, I figured there were no fences all the way back to Arizona.

I began tracking them, my thirst mounting. After four or five miles, I saw a flash of white in a stand of piñon trees about a mile away. I crept up within several hundred yards, and then filled my hands with dirt, holding them out and whistling as if I had a handful of oats. Pedernal jerked his head up and immediately took off at a gallop, with the other horses following. I followed them for hours in the blinding heat, but they wouldn't let me get nearer than several hundred yards. The horses eventually entered a country of badlands west of Dowa Y'alanne, and I caught them when they stopped to drink at a mudhole.

Walter and Engwahela finally arrived at the camp late in the afternoon. He had walked back to the pueblo, where he had enlisted Edward's help in tracking the horse. Walter had been tracking her on foot about 16 miles from our camp, carrying a bucket of oats, when the horse came up from behind and nudged him with her muzzle.

A week of riding brought us past El Morro and across the Zuni Mountains to the western edge of El Malpais, the Bad Country: a vast basin filled with black, twisted lava. Many Coronado scholars have assumed that Alvarado and his Pecos guides went around the Malpais. No horse or mule, they say, could have crossed it; and besides, they claimed, there was no water in the lava beds.

I did not agree. I thought it unlikely that the Indians leading Alvarado would have taken a 30-mile detour just to avoid a little rough, waterless country. I wanted to prove a horse could be ridden across the Malpais.

We started into the lava the morning of August 13. We had chosen to cross at a seven-mile stretch that the National Park Service had indicated was the Dominguez-Escalante Trail, an old Spanish route. In a half-mile we found ourselves in the midst of the strangest landscape in North America, surrounded by great rafts and blocks of black lava. We decided to scout ahead on foot.

Walter volunteered, and returned an hour later, his face white. "That's the closest I've ever come to seeing hell itself," he said, gasping. The bottoms and sides of his boots were scored by the lava, and his hands were cut. We turned back.

We rode along the edge of the Malpais and that evening came to a ranch house. The rancher, a man named Alfredo Mirabal, listened to our story. "That ain't the Dominguez-Escalante Trail," he said, laughing. "That's just something made up by the park people. That ain't even a trail at all. The *real* trail, the old Indian trail, starts right here on my property. They wanted to mark it but I told 'em to get the hell out."

We got into his four-by-four pickup and bumped across the prairie alongside the black wall of lava, and he showed us the trailhead. It was so well hidden that I didn't see it until I was five feet from it. It was just wide enough for a horse, and it was clearly man-made.

"The Indians used to escape from the cavalry in there," Alfredo said. "The cavalry could only follow 'em one horse at a time, and the Indians would just pick 'em off." When I asked about water, he snorted. "There's more water in that Malpais than anywhere else around here. You just gotta know where to find it." He then showed us yet another trail across the lava, closer to where we had decided to camp. This trail, which we crossed the next day, was so wide and flat you could have driven a jeep along it.

The distance from Acoma to Albuquerque was about 75 miles. In 60 of those miles we found water in only two places: a chocolate-milk-like substance in a mud wallow and a green scummy brew in an abandoned sheep corral. On the third day, with our supply running out again, we saw a windmill and cattle tank in the distance. Cattle were clustered around it, meaning there had to be water. We broke into a gallop. I arrived first, to find a beautiful tank brimming with water.

When I rode around it, my horse suddenly shied away. There, lying mostly submerged in the water trough, was a cow whose head had been partly blown away by a very recent shotgun blast (this method of settling disputes is common in these parts).

"We can't drink this water," I said.

Walter got off his horse and peered at the cow. The cow's eyes rolled and her legs trembled. He looked back at me. "I don't know why not, the cow ain't dead yet."

"You have a point there," I said, and we proceeded to fill up our canteens. We would drink that water all the way to Albuquerque.

Every day the hot blue outline of Sandía Crest had been rising a little higher over the horizon, and every night the glow of Albuquerque became stronger. On the fourth day we came to the edge of an escarpment. Spread out below us in the broad valley along the Rio Grande was Albuquerque itself, cloaked in a pool of brown smog. We wound our way to the edge of the metropolis through piles of garbage, burned mattresses, discarded refrigerators and dead animals, and then rode straight down a busy avenue into the heart of the city.

Alvarado and his guides probably struck the Rio Grande very close to this spot. What he saw were emerald fields unfurling along the floodplains of the river, and dozens of large, prosperous pueblos lining the terraces above the riverbanks, a "kingdom" the Indians called Tiguex (present-day Keresan). Alvarado sent back word to Coronado that this rich valley would be a good place for the army to spend the winter. He continued up the river and then went east at Galisteo Creek.

Seventy miles up the Rio Grande we rode into Santo Domingo Pueblo, at the confluence of Galisteo Creek. We were given a warm welcome by pueblo governor Ernie Lovato, a short man with a powerful chest and a rapid, intense way of talking, who fed us a splendid traditional dinner of mutton stew and Indian bread. "Many centuries ago," he pointed out, "we greeted Coronado the same way. There is a deep tradition of hospitality among the Santo Domingo people. I hope," he continued with a wry smile, "that you do not return our hospitality the way Coronado did."

Alvarado arrived at Pecos in late September. The powerful pueblo sat on a tongue of rock in the middle of a green valley, with the Sangre de Cristo Mountains rising up behind it. The Indians streamed out and greeted Alvarado with skirling bone flageolots and beating drums, and loaded him down with woven blankets, buttery-soft animal skins, turkeys, jerked beef and rich turquoises. The Pecos were "feared throughout that country," one Spanish soldier wrote; and they "boast that no one has been able to conquer them."

HERE, AT PECOS Pueblo, the ultimate fate of Coronado's expedition was sealed. Incessantly inquiring after gold, Alvarado came across a fast-talking Plains Indian slave at the pueblo, whom they nicknamed the Turk. The Turk spun a marvelous story about a kingdom called Quivira. Why, said the Turk, in Quivira everyone ate from plates of silver and drank from jugs of gold, while the king napped under a tree festooned with gold bells that tinkled in the wind.

Once again the Spanish felt the fever of gold upon them, and decided to seek Quivira the following spring. Meanwhile, Coronado and his army had arrived at the Rio Grande for the winter. In return for the hospitality of the Tiguex Indians, who had welcomed the Spanish with gifts and food, the Spanish seized a pueblo and went around confiscating clothing and blankets. One pueblo revolted, but when the Indians surrendered under a promise of clemency the Spanish burned them at the stake anyway. Other pueblos rose up, and by spring much of the green Rio Grande valley had been laid waste by the conquerors.

The following April the army decamped and went in search of Quivira, venturing 1,000 miles into the Texas Panhandle. Coronado and a handful of men pushed on as far as central Kanas, where they entered the legendary kingdom: it consisted of a few dozen straw huts along the Arkansas River. The Turk was promptly garroted and Coronado returned with his army to Mexico, his expedition a failure, his fortune gone and his reputation destroyed. He died in obscurity at age 44.

Without a calendar, we had lost track of the days, and I am still not sure exactly when we finally rode into Pecos Pueblo, our final destination. It was nearly September, and the nights had already begun sharply cooling off. As we rode into the parking lot of the Pecos National Monument, thunderheads piled up over the Sangre de Cristo Mountains, sending down gunmetal-blue columns of rain. The great pueblo, where Alvarado had been greeted by flutes and drums, lay in ruins, having been abandoned in 1838 after a long decline. Today it is a popular tourist site.

Walter and I unloaded our packhorses in front of a modern visitor's center, and then we rode up to ruins on an asphalt service road. We surprised a group of tourists in pastel pantsuits and straw hats, who converged on us and began taking pictures and asking questions. Then the wind picked up and blew a cloud of dust through the ruins, which was followed by a scattering of heavy drops. The crowd, clutching their hats and cameras, went running for their cars.

Walter and I sat on our horses and watched the eroded walls of the ancient pueblo, once the greatest city in America north of Mexico, darken in the rain.

"I guess it's over," I said after a moment.

"Yup," said Walter. "Let's go home."

UPDATE

After this article was published, I wrote a nonfiction book about the journey, titled Cities of Gold, *which was published by Simon & Schuster and is still in print.*

That thousand-mile journey, now thirty-five years distant, led, in a curious way, to the nonfiction book I am writing now, titled To the Last Man: The Pleasant Valley War and the Creation of the Western Myth. *It came about like this: The remotest section of our trail in Arizona ran from the Salt River to the top of the Mogollon Rim, through more than a hundred miles of rugged mountains. This was the region discussed in the article, which Coronado called the* despoblado, *the "uninhabited wilderness,"* which almost killed the Spanish—and just about did us in too. In the middle of the despoblado, we rode unexpectedly into a tiny town called Young, situated in a mile-high grassy valley and deeply isolated from the outside world. Young was one of the remotest towns in the lower forty-eight, with the nearest supermarket at the time sixty miles away, thirty of those over bone-shaking dirt roads. At the time of our visit, it was inhabited by a few hundred ranching and mining families.

Walter and I put up at a ranch run by an old-timer named Tobe Haught, whose family had been among the earliest settlers in the valley. That evening, Tobe got to telling stories, and he recounted an extraordinary episode in the history of Young, one in which his family played a part. The town, he said, used to be called Pleasant Valley, and it had been the site of the most violent range war and feud in American history. Between 1882 and 1892, thirty to fifty people died in gunfights, murders, vigilante hangings, or simply disappeared—approximately half the male population of the valley. All the men in two extended families, the Tewksburys and the Grahams, were exterminated until only two were left: Edwin Tewksbury and Tom Graham. Years later, when everyone thought the war was over, Tewksbury murdered Graham in a spectacular act that made national news, followed by several sensational trials that filled newspaper columns for three years.

That final killing in the war assumed mythic status in Western history. It was the basis for Zane Grey's 1921 novel *To the Last Man*, from which I've shamelessly taken the title of my book, as well as three of the earliest Western movies, which introduced a number of iconic themes into our national consciousness. What makes the Pleasant Valley War so central to the history of the West is that it was much more than a sordid feud or a conflict over grazing. In that fatal decade, many of the dark and powerful forces that went into the making of the American West and its myths—good and bad—came together in this remote and beautiful valley, with explosive results. The Pleasant Valley War was the ur-source for such tropes as sheepmen vs. cattlemen, evil New York railroad barons clearing the land with hired killers, spectacular shootouts, brutal family feuds, and the violent, genocidal removal of Native peoples from their land. The war also revealed the ugly treatment of so-called "half-breeds" in nineteenth-century America, of which the Tewksburys were a conspicuous example.

In 1887, the Pleasant Valley War broke into the national news. Tewksbury's final killing of Graham five years later—and how he used a relay of superb horses to traverse hundreds of miles of mountains, do the deed, and fly back to his mountain stronghold—was the talk of the nation. The war aroused the concern of Congress, which concluded that the Arizona Territory was ungovernable and undeserving of being admitted as a state. Since then, even as the myths inspired by the war continue to haunt our books and movies, the original story faded into obscurity until it was forgotten. In To the Last Man, *I hope to bring back into the American consciousness this epic story of the West in all its glory, terror, violence, ugliness, and fascination.*

ACKNOWLEDGMENTS

I OWE A great debt to Otto Penzler, to whom this book is dedicated, for suggesting the idea for it in the first place. I also would like to thank the many magazine editors I worked with on these pieces, who contributed a great deal to making them what they are. This includes, most particularly, the late John Bennet, who gave me my first assignment in the *New Yorker*, as well as David Remnick, Tina Brown, Dorothy Wickenden, David Kuhn, Lewis Lapham, Leo J. Carey, Deirdre Foley-Mendelssohn, Alan Burdick, Anthony Lydgate, Sharon DeLano, Joy de Menil, Don Moser, and the late Alan Ternes. I'm also indebted to the many excellent fact checkers and copy editors, too numerous to name, who saved me from grammatical and factual embarrassment.

ABOUT THE AUTHOR

DOUGLAS PRESTON worked as a writer and editor for the American Museum of Natural History and taught writing at Princeton University. He has written for the *New Yorker, Natural History, National Geographic, Harper's, Smithsonian,* and *The Atlantic.* The author of several acclaimed nonfiction books—including *The Lost City of the Monkey God, Cities of Gold,* and *The Monster of Florence*—Preston is also the coauthor, with Lincoln Child, of the bestselling series of novels featuring FBI Agent Pendergast.